NATIONAL ACADEMIES Sciences Engineering Medicine

NATIONAL ACADEMIES PRESS
Washington, DC

Investing in Resilient Infrastructure in the Gulf of Mexico

Micah Lowenthal, *Editor*

Erin Mohres, *Rapporteur*

Office of Special Projects

Policy and Global Affairs

Gulf Research Program

Proceedings of a Workshop

THE NATIONAL ACADEMIES PRESS 500 Fifth Street, NW Washington, DC 20001

This activity was supported by the Gulf Research Program of the National Academies of Sciences, Engineering, and Medicine. Any opinions, findings, conclusions, or recommendations expressed in this publication do not necessarily reflect the views of any organization or agency that provided support for the project.

International Standard Book Number-13: 978-0-309-68847-5
International Standard Book Number-10: 0-309-68847-7
Digital Object Identifier: https://doi.org/10.17226/26559

This publication is available from the National Academies Press, 500 Fifth Street, NW, Keck 360, Washington, DC 20001; (800) 624-6242 or (202) 334-3313; http://www.nap.edu.

Copyright 2022 by the National Academy of Sciences. National Academies of Sciences, Engineering, and Medicine and National Academies Press and the graphical logos for each are all trademarks of the National Academy of Sciences. All rights reserved.

Printed in the United States of America

Suggested citation: National Academies of Sciences, Engineering, and Medicine. 2022. *Investing in Resilient Infrastructure in the Gulf of Mexico: Proceedings of a Workshop*. Washington, DC: The National Academies Press. https://doi.org/10.17226/26559.

The **National Academy of Sciences** was established in 1863 by an Act of Congress, signed by President Lincoln, as a private, nongovernmental institution to advise the nation on issues related to science and technology. Members are elected by their peers for outstanding contributions to research. Dr. Marcia McNutt is president.

The **National Academy of Engineering** was established in 1964 under the charter of the National Academy of Sciences to bring the practices of engineering to advising the nation. Members are elected by their peers for extraordinary contributions to engineering. Dr. John L. Anderson is president.

The **National Academy of Medicine** (formerly the Institute of Medicine) was established in 1970 under the charter of the National Academy of Sciences to advise the nation on medical and health issues. Members are elected by their peers for distinguished contributions to medicine and health. Dr. Victor J. Dzau is president.

The three Academies work together as the **National Academies of Sciences, Engineering, and Medicine** to provide independent, objective analysis and advice to the nation and conduct other activities to solve complex problems and inform public policy decisions. The National Academies also encourage education and research, recognize outstanding contributions to knowledge, and increase public understanding in matters of science, engineering, and medicine.

Learn more about the National Academies of Sciences, Engineering, and Medicine at **www.nationalacademies.org**.

Consensus Study Reports published by the National Academies of Sciences, Engineering, and Medicine document the evidence-based consensus on the study's statement of task by an authoring committee of experts. Reports typically include findings, conclusions, and recommendations based on information gathered by the committee and the committee's deliberations. Each report has been subjected to a rigorous and independent peer-review process and it represents the position of the National Academies on the statement of task.

Proceedings published by the National Academies of Sciences, Engineering, and Medicine chronicle the presentations and discussions at a workshop, symposium, or other event convened by the National Academies. The statements and opinions contained in proceedings are those of the participants and are not endorsed by other participants, the planning committee, or the National Academies.

Rapid Expert Consultations published by the National Academies of Sciences, Engineering, and Medicine are authored by subject-matter experts on narrowly focused topics that can be supported by a body of evidence. The discussions contained in rapid expert consultations are considered those of the authors and do not contain policy recommendations. Rapid expert consultations are reviewed by the institution before release.

For information about other products and activities of the National Academies, please visit www.nationalacademies.org/about/whatwedo.

PLANNING COMMITTEE FOR INVESTING IN RESILIENT INFRASTRUCTURE IN THE GULF OF MEXICO

Members

DAVID E. DANIEL (Chair), University of Texas
THOMAS BOSTICK, Bostick Global Strategies
M. GRANGER MORGAN, Carnegie Mellon University
SARA ORTWEIN, XTO Energy, Exxon Mobil Corporation (retired)

Project Staff

MICAH LOWENTHAL, Senior Program Director
MONICA STARNES, Senior Program Officer
MEGHA KHADKA, Research Associate
TERI THOROWGOOD, Executive Assistant

CNA Workshop Designers and Facilitators

ERIN MOHRES, Project Lead and Lead Facilitator
SEBASTIAN BAE, Workshop Controller
ANGIE DE GROOT, Facilitator
ELEANORE DOUGLAS, Lead Analyst
MARK ROBERTS, Facilitator
YEE SAN SU, Advisor
RIDDHI SUVA, Facilitator

ACKNOWLEDGMENTS

This Proceedings of a Workshop was reviewed in draft form by individuals chosen for their diverse perspectives and technical expertise. The purpose of this independent review is to provide candid and critical comments that will assist the National Academies of Sciences, Engineering, and Medicine in making each published report as sound as possible and to ensure that it meets the institutional standards for quality, objectivity, evidence, and responsiveness to the study charge. The review comments and draft manuscript remain confidential to protect the integrity of the deliberative process.

We thank the following individuals for their review of this proceedings: Maria Honeycutt, The White House; Amanda Martin, State of North Carolina; David Owens, Da'VAS LLC; Hanadi Rifai, University of Houston; Charles Williams, Center for Offshore Safety (ret.); Roy Wright, Insurance Institute for Business and Home Safety.

Although the reviewers listed above provided many constructive comments and suggestions, they were not asked to endorse the content of the proceedings nor did they see the final draft before its release. The review of this proceedings was overseen by Gerald Galloway (NAE), University of Maryland, College Park. He was responsible for making certain that an independent examination of this proceedings was carried out in accordance with the standards of the National Academies and that all review comments were carefully considered. Responsibility for the final content rests entirely with the rapporteurs and the National Academies.

PREFACE

The United States is embarking on a surge in investment in maintaining, improving, and replacing infrastructure. Some infrastructure grant programs will see their budgets increase by a factor of 10. These increases are temporary, and even when they are in effect they will be insufficient to fund all of the valuable infrastructure projects that would benefit the nation, so some prioritization will have to be done. These investments present opportunities, which may benefit or harm the lives of our people for the coming decades and beyond. To reap the benefits and avoid the harms, we need processes for informing those prioritization decisions with science, engineering, community involvement, and systems-level thinking that address needs that are changing with our changing environment and uses. As National Academy of Sciences' President Marcia McNutt noted, the decisions and investments done right will save lives and figuratively pave the way to a more prosperous future.

Many of the federal programs that will receive major infusions of funds have mature processes for deciding among projects and proposals, but they generally only assess options within one sector and only rarely do they factor in the bigger picture of planning for different types of infrastructure or different societal functions or benefits. Because they anticipated this need, the National Academies of Sciences, Engineering, and Medicine developed an initiative on infrastructure investment prioritization and decided to begin with an interactive workshop on investing in resilient infrastructure in the Gulf of Mexico region. For this workshop, we convened topical experts, federal managers from multiple agencies, members of the affected communities, state governments, industry, and experts on the science and technology of infrastructure and of the stressors that we expect our infrastructure to face.

Typical workshops are more like symposia, with a series of talks and questions directed to the speakers. This infrastructure workshop was totally unlike those. The workshop began with two stage-setting talks, one from White House National Security Council Director for Resilience and Response Jason Tama, and one from former Commanding General of the U.S. Army Corps of Engineers Tom Bostick. Thereafter, CNA, a nonprofit research and analysis organization, facilitated discussions working through scenarios that the National Academies and CNA designed to explore infrastructure issues in the Gulf region. Because this was not a typical workshop, this summary does not look like a typical National Academies workshop proceedings; a sequential recounting of comments would not as accurately reflect the nature of the workshop as does this synthesis of the remarks, actions, and discussions. This also means that where the text says "Participants said that party X should do Y," it does not necessarily reflect the consensus of all the participants and it is not a recommendation of the National Academies.

The National Academies are very pleased with the success of the Investing in Resilient Infrastructure in the Gulf of Mexico Workshop. It aligned with the National Academies and the Gulf Research Program's (GRP) broader vision to support a safer, more resilient, and sustainable future for the Gulf and all those who call the region home, using science, engineering, and medical knowledge to empower its citizens and to enhance Gulf offshore energy safety, environmental protection and stewardship, and health and resilience. By demonstrating a structured process to identify valuable projects and develop a framework to prioritize investments and seeking to harmonize national, regional, and local interests, this workshop took an important step toward supporting a more resilient Gulf region.

The Planning Committee for Investing in Resilient Infrastructure in the Gulf of Mexico Workshop provided essential guidance and feedback in the development of the workshop. The committee members minimized their participation on the group discussions to ensure that the organizers elicited views and ideas from the invitees rather than the planners. Erin Mohres and her team at CNA did an impressive job developing approaches to help participants work together on the issues that the National Academies identified and to support the overall institution's and the GRP's missions and goals, and they ran an excellent event from facilitation to materials. In particular, in addition to Ms. Mohres, Sebastian Bae, Angie De Groot, Eleanore Douglas (Dr. Douglas's description of the decision framework she put together for this effort is in Appendix C), Mark Roberts, Yee San Su, and Riddhi Suva all made important contributions to the design and implementation of the workshop, and consequently to this proceedings. Monica Starnes, Megha Khadka, and Teri Thorowgood managed the overall project, including input and guidance on substantive content as well as logistical support. All of us foremost are grateful to the participants who brought their expertise, their enthusiasm, and their sense of purpose to this demonstration activity, grappling with engineering, social equity, interoperability, and changing climate.

This workshop should be just the first step in an exciting initiative. Please look out for future activities.

Micah Lowenthal
Director, Office of Special Projects
National Academies of Sciences, Engineering, and Medicine

CONTENTS

SUMMARY 1

PART 1 INTRODUCTION 9
    Proceedings Organization 10
    Convening a Workshop 11
    Setting the Context Through Keynotes 13
    The Exercises 16
    The Scenarios 21
    The Hybrid Format 24

PART 2 PROJECT IDENTIFICATION AND PRIORITIZATION 25
    Observations and Key Takeaways: Hurricane Scenario 27
    Observations and Key Takeaways: Protracted Oil Spill Scenario 30
    High-Priority Project and Investment Ideas 33

PART 3 PRIORITIZATION FRAMEWORK 37
    The Prioritization Criteria 37
    Relative Weighting of the Macro-Criteria 38
    Elaboration on the Sub-Criteria 39

PART 4 NEXT STEPS 53

APPENDIXES

A TAKEAWAYS AND OBSERVATIONS BY DOMAIN 55

B COMPLETE LIST OF PROJECT IDEAS IDENTIFIED 65

C PRIORITIZATION FRAMEWORK: RESEARCH AND RATIONALE 77

D EVENT AGENDA 83

E BIOGRAPHICAL SKETCHES OF SPEAKERS AND STEERING COMMITTEE 87

F EVENT PARTICIPANTS 91

Summary

To help prioritize among possible investments to improve the resilience of built infrastructure in the Gulf of Mexico region,[1] the National Academies of Sciences, Engineering, and Medicine convened a diverse group of experts for a 3-day interactive workshop on November 15, 16, and 18, 2021. This workshop was held as communities surrounding the Gulf continue to experience frequent, destructive disasters, some infrastructure in the region continues to degrade or fail from exceeded capacity and delayed maintenance and replacement, and climate change threatens previously unimagined impacts. On the same day that the workshop began, President Biden signed the bipartisan Infrastructure Investment and Jobs Act (IIJA, P.L. 117–58), which will offer opportunities to meet identified needs with additional funds. But even with the IIJA, the funds are limited, so the region and the nation need ways to prioritize investments informed by sound science and engineering.

The workshop, titled Investing in Resilient Infrastructure in the Gulf of Mexico, demonstrated and refined a process to help inform recommendations for prioritizing infrastructure investments across sectors and anchored in the Gulf region energy industry. The workshop had four main objectives:

1. Use two scenarios—a hurricane and a protracted oil spill—to start discussions on infrastructure vulnerabilities
2. Identify and prioritize projects or investments to reduce those vulnerabilities, thereby strengthening infrastructure resilience in the Gulf region
3. Identify criteria that would drive infrastructure investment decisions and prioritization
4. Define criteria that will form the foundation of a repeatable prioritization framework for infrastructure investment decisions through cross-comparison of insights from each of the two scenarios

The workshop also introduced three policy questions to guide discussion:

1. What choices would maximize resilience return on federal investments in infrastructure and promote equity, fairness, economic effectiveness, and other goals and requirements?
2. How can we overcome obstacles to making the best investments in infrastructure?
3. How can we mitigate against private infrastructure assets becoming public liabilities?

As a demonstration activity, this workshop does not provide a definitive or authoritative prioritized list of projects, nor does it provide a consensus of the workshop participants. Rather, the workshop provides a first step in a process bringing together those with responsibilities and

---

[1] There are many different definitions of infrastructure in use today. This workshop focused on the built environment, including human-fabricated physical structures and interventions in the natural environment. Social infrastructure is supported by these investments, and although it may require its own major investment, that was outside of the scope of the workshop.

stakes in prioritization across sectors, jurisdictions, and interests. This summary provides insights gained from the process and innovative ideas that arose in discussions.

WORKSHOP OVERVIEW

Approximately 50 infrastructure, oil and gas, petrochemical, emergency management, and Gulf region experts from federal, state, and local governments, industry, non-governmental organizations, and academia participated in this 3-day workshop. Half of the participants joined Day 1 on November 15 for a set of in-person exercises focused on a hurricane scenario; half of the participants joined Day 2 on November 16 for a set of in-person exercises focused on a protracted oil spill scenario. The groups were split to ensure that they were small enough for every attendee to participate substantively and that appropriate expertise was represented in each group. Exercise materials included detailed scenario descriptions and chains of adverse events likely prompted by each scenario. Participants reviewed and updated that information, discussed vulnerabilities related to each scenario that impact local infrastructure, and then brainstormed and prioritized projects to address those vulnerabilities. As used here, "project" refers to any concrete action that can be taken to prevent or mitigate any of the adverse events prompted by the scenarios. All participants joined Day 3 virtually on November 18 for a set of capstone exercises that compared and built upon ideas discussed in Days 1 and 2. (See Figure S-1 for an overview of the workshop.)

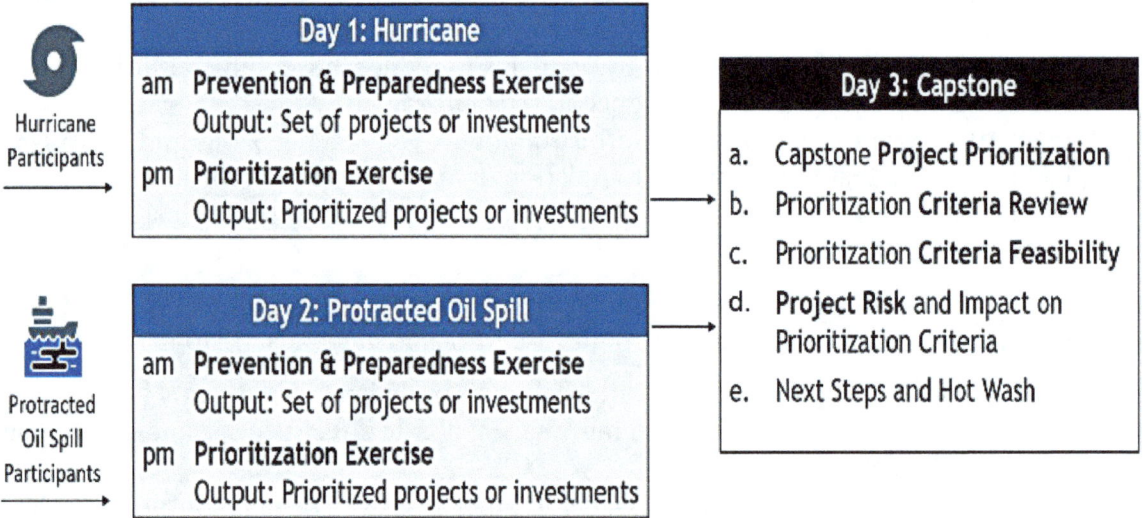

FIGURE S-1 Overview of workshop plan.

Discussion and information collection were structured around four domains (see Figure S-2): (1) oil, gas, and petrochemical industry functions; (2) other infrastructure functions; (3) society's needs; and (4) environmental protection. These domains provided a useful mechanism to organize inputs and feedback. On Days 1 and 2, participants joined small groups designed to include experts with diverse backgrounds, and all participants rotated through all domains. On Day 3, participants joined small groups designed to include experts with similar backgrounds, and they spent their time in a domain that closely aligned with that background.

Petrochemical Industry Functions
- Production (Crude Oil)
- Refining
- Transport
- Retail

Society's Needs
- Housing
- Education
- Healthcare
- Commerce

Other Infrastructure Functions
- Communications
- Electricity
- Water
- Transportation

Environmental Protection
- Air
- Land
- Water

FIGURE S-2 Domains around which workshop discussion and information collection were organized.

The National Academies also arranged for three keynote speakers to set the context for this event. Jason Tama, White House National Security Council Director for Resilience and Response, emphasized what an opportune time it was for this workshop in light of the signing of the bipartisan Infrastructure Investment and Jobs Act into law. He hoped conference participants would identify key principles to prioritize investment in infrastructure resilience and then share those principles with decision makers. Lt. General Thomas P. Bostick, retired Commander of the U.S. Army Corps of Engineers, pushed for broad, systems-level thinking to inform infrastructure investment. He stressed the importance of developing ideas that are measurable and actionable, and he mentioned stakeholder involvement and engagement in infrastructure investments as absolutely imperative. Finally, Marcia McNutt, President of the National Academy of Sciences (NAS), spoke about setting priorities on upgrading infrastructure, striving for equity in outcomes, and harmonizing stakeholder interests and values. She advocated for forward-thinking investments that consider future changes such as climate change.

WORKSHOP TAKEAWAYS

The Investing in Resilient Infrastructure in the Gulf of Mexico Workshop demonstrated and advanced a process and framework that federal agencies and their partners in other jurisdictions might employ to identify and prioritize investments that would increase infrastructure resilience. Expert participants from the public sector, private sector, and academia emphasized the necessity of community engagement, of working to overcome long-standing obstacles, and of moving into the future, if infrastructure resilience is to be achieved. Key workshop-wide takeaways fell into these three broad categories and are outlined in Box S-1 and described after the box. They are expanded upon throughout the body of this proceedings. Workshop participants also identified criteria that form the basis of a prioritization framework for selecting high-value projects that would likely increase infrastructure resilience in the Gulf of Mexico. Finally, participants identified 306 individual project ideas; projects identified as high priority are included in the body of this proceedings, and a comprehensive list is located in Appendix B.

> **BOX S-1 Overview of Workshop-wide Key Takeaways**
>
> **Necessity of Community Engagement**
>
> - Emphasizing local leadership
> - Promoting community involvement
> - Improving communications and transparency
>
> **Call to Overcome Long-standing Obstacles**
>
> - Moving from survivability to thrivability
> - Focusing on prevention in addition to response
> - Incorporating resilience and equity consistently
>
> **Need to Move into the Future**
>
> - Building up our data and analysis
> - Embracing technology and modernization
> - Balancing regulations versus incentives carefully

The following descriptions elaborate upon the key takeaways outlined in Box S-1, highlighting points made by some participants across all 3 days of the workshop. Note that these do not necessarily reflect a consensus of the participants' perspectives.

Necessity of Community Engagement

Emphasizing Local Leadership

Participants said that local leadership is essential for projects to be effective, ensuring that local needs are understood, communicated, and considered; buy-in of new projects is achieved; and project outcomes benefit all relevant stakeholders. This includes increasing local leadership in disaster response decision-making, which is critical to increasing resilience. Many of these decisions are too often made at the state or federal levels, insufficiently including local responders.

Promoting Community Involvement

Decision makers and project leads are more likely to achieve stakeholder buy-in if they engage a community early and often.[2] Such engagement includes getting input from lower-income and impoverished communities on project approaches, partnerships between project managers and the community, and involving local leadership in project-related planning and execution. Community engagement helps to ensure that community members understand the benefits that new projects or changes can offer. Community engagement should not be thought of as a one-time event, a final output, or a stage, but as an ongoing process that is vital to building and maintaining trust.

---

[2] Community engagement pertains to local government as well as the residents, businesses, and organizations in that locality.

Improving Communications and Transparency

Effective communications between government program managers and the community would be ongoing, frequent, and clear. Such communications would include more educational information about projects and their benefits, and also information about local threats and hazards, what kinds of impacts the community should expect, how to prepare for those impacts, and how to get support. Participants said that the media must be a partner. A strengthened relationship between the media and all levels of government, and enhanced information sharing between them during a protracted disaster response, especially an oil spill response, should reduce the broadcasting of misinformation.

Overcoming Long-Standing Obstacles

Moving from Survivability to Thrivability

Basic survival is still at risk post-emergency in the Gulf region. Some participants said that, first and foremost, a key goal of projects must be to strengthen the availability of basic human needs, including drinking water, wastewater systems, and housing. Projects should support rapid restoration of power and communications to promote a return to stability. But through these investments, it is also critical to consider what infrastructure will improve day-to-day quality of life, reduce the likelihood of failures (see the next takeaway), and improve equity of outcomes when there are failures of infrastructure.

Focusing on Prevention in Addition to Response

Workshop participants expressed aspirations that utilities not fail under anticipated stresses; that potential contaminants not affect drinking water; and that people, residences, and infrastructure not be located in areas designated as high risk for flooding or other hazards. It is critical to continuously maintain response capability, but increasing attention will have to be paid to prevention of adverse impacts from infrastructure failures happening to begin with. Examples of preventive measures included undergrounding power lines, phasing out tall communications towers, and increasing robustness and resilience of petroleum storage tanks. The region's land use plans and regulations affect resilience, and reexamining them, moving infrastructure out of high-risk zones, could have major positive effects, although this is a question of will and not of ability.

Incorporating Resilience and Equity Consistently

Participants emphasized the need to not only assess new investments specifically for their consideration of resilience and equity but also to build them into every activity taken on. Resilience is a mindset that focuses on always becoming more capable of withstanding stresses, including disasters. Enhancing all phases of emergency management is critical to increasing resilience—projects that focus on improving prevention, preparedness, planning, response, and recovery are all worth considering. Equity must also become a constant consideration. Not only are marginalized and disadvantaged people the most vulnerable to failures of infrastructure, they are the most vulnerable to infrastructure investments that disregard their welfare. To ensure that new investments do not aggravate inequities, but rather work to address and resolve them, funders can use more environmental, social, and governance mandates for corporations and community leadership in new projects.

Moving into the Future

Too many projects and programs are stuck in the past, focused on fixing or restoring outdated infrastructure designed for environments that have changed. The following takeaways focus on enabling communities to meet future needs.

Building up Data and Analysis

Effective prioritization of projects and investments requires both baseline data for an improved understanding of current conditions (environmental, social, constructed) and to better monitor and measure new projects' effectiveness. Decision makers also need more long-term studies regarding the specific impacts that can be expected from threats, hazards, changing use patterns, complex disasters, and climate change. Baseline data needs to include the specific locations of existing infrastructure systems, health and economic information, and the natural environment. Decision makers and communities need data collection to be an ongoing activity, conducted continuously or at regular intervals. Building on these data, long-term studies should include predictive modeling, participants said.

Embracing Technology and Modernization

Participants reported that the Gulf region is behind in its adoption of modern infrastructure technology. Better and broader use of sensors across the board would increase situational awareness, provide data, and protect workers—monitoring roadways for flooding, water for contaminants, fuel tanks for damage, and carbon monoxide emission during generator use are just a few examples. Adopting more remote-sensing and autonomous technologies such as the use of drones for damage assessment arose as priorities. Better systems are needed to monitor the power grid and the thousands of miles of pipeline in the Gulf. New oil spill barrier construction technology and environmentally friendly biodegradable dissolvents and surfactants were highlighted. There is also a need for more investment in research and development to continuously improve and advance infrastructure and the operation, monitoring, and maintenance associated with it.

Considering Regulations versus Incentives Carefully

Achieving resilience requires a new balance of regulation and incentives. Although using incentives was seen as valid in some cases, incentives have not always worked, and so some participants said that regulations, along with increased enforcement power of several federal agencies to ensure those regulations are followed, should be adopted. When incentivizing, stricter monitoring is needed to ensure those incentives are prompting the action and outcomes that are being sought.

PRIORITIZATION CRITERIA

Figure S-3 depicts the decision-making criteria identified by participants during this workshop as particularly useful and relevant to help prioritize investments that will increase infrastructure resilience in the Gulf region. Although derived in the specific context of infrastructure resilience in the Gulf region, these criteria and the categories that group them provide the foundation of a framework that can be adopted and adapted by agencies and other organizations faced with making similar investment decisions. More details about these criteria can be found in the body of this proceedings.

Environment	Economy	Society	Resilience	Project Governance
• Advancement of scientific understanding of the environment • **Biodiversity** • **Climate change** • Degradation prevention • **Do no harm / reduce harm to environment** • **Land use** • Maintenance • Mitigation • Persistence of efforts • **Riverine and coastal erosion** • Safe living and working spaces • Sustainability • Understanding of environment baselines and changes • **Water/air quality**	• Business continuity • **Commerce support** • Continuity/recovery of economic activity • Cost-effectiveness • **Economic development** • **Employment** • Equity • **Financial need** • **Individual support** • Labor rights and working conditions • Multiple impacts • Proximity to major economic and/or residential centers • Spending drivers • **Tax base**	• **Access to information / awareness** • Benefits go to underserved communities • **Community engagement** • Connections • Cultural value • **Education** • **Equity and inclusion** • **Health and welfare** • **Public safety/order** • Public trust • Return to stable society/normalcy • Social and political acceptability • Survivability (e.g., water, basic needs)	• **Adaptability** • Expedited recovery • **Exposure reduction** • Focus on preparedness and response • **Hardening of infrastructure** • Info management and comms • **Interdependence with other systems** • Long-term impact • Planning and prep (prevention focus) • Rapid deployment for response • **Redundancy** • **Supportive of the system / network** • Understanding AND action	• Alignment with existing plan • Accounting for cyber/physical security implications • **Community inclusion** • **Data and information sharing** • **Feasibility** • Measurable • **Outcomes focus— effectiveness, return on investment (ROI)** • **Oversight and compliance** • Project readiness • Repeatability / scalability • **Stakeholder coordination** • **Technical merit**

FIGURE S-3 Criteria that support a prioritization framework for infrastructure projects in the Gulf region.
NOTE: Boldface indicates criteria identified as significant for infrastructure projects in the Gulf region.

In the closing session, participants expressed how valuable they found the workshop, bringing together such a diverse group to share and learn from each other, and they made suggestions for improvements and next steps. On that last day of the workshop, NAS President Marcia McNutt said that the National Academies plan for the workshop to be just the first step in a big, bold, and exciting effort to help the United States make good use of limited funds to meet future infrastructure needs. She thanked the participants and expressed the hope that their organizations and the National Academies can work together on this critical need for the nation.

Part 1

Introduction

States along the Gulf of Mexico experience some of the most frequent and destructive natural disasters in the country, including flooding, drought, tornadoes, and hurricanes. As impacts from climate change and more intense weather events affect the region, the increasing toll that it takes on infrastructure, and in turn on the people of the Gulf region, cannot be ignored. Record-breaking storms and weather events are occurring more frequently, and the costs associated with repairing the damage and rebuilding towns and cities has steadily increased. In 2020 alone, the United States faced 22 extreme weather events that each cost the nation more than $1 billion.[1] In the face of new realities of extreme weather events and the need to adapt to a shifting normal due to climate change, the nation must look for new ways to invest in society's infrastructure systems so that we not only rebuild from disasters but also systematically incorporate the elements of prevention, mitigation, and resilience into planning efforts.

The current state of infrastructure in the United States is in poor and failing condition.[2] The Gulf states, which produce, refine, and transport much of the nation's oil and gas, are no exception. The roads, bridges, inland waterways, and ports that transport goods from abroad to every corner of the country are in disrepair. Hurricanes, combined with storm surge, rising sea levels, and record rainfall have damaged infrastructure beyond roads and bridges. Water distribution systems, communication networks, and power generation and distribution systems are too frequently knocked offline by these and other weather events and cause serious suffering and harm. Communities, small businesses, hospitals, and schools have been destroyed or deemed uninhabitable due to storm-induced flooding. And oil spills, both the catastrophic disasters that make front-page news and lesser-known oil spills that can last for years or decades, continue to release toxic chemicals from privately held wells and pipelines, harming the environment and people's health.

The bipartisan Infrastructure Investment and Jobs Act (IIJA, P.L. 117-58), which has the potential to address long-overdue improvements to our roads, bridges, railways, and ports,[3] provides a generational opportunity to fix not only dilapidated infrastructure but also the way in which we prioritize the projects that we fund. Currently, the federal government aids in rebuilding existing infrastructure, often in ways that do not prevent future infrastructure failures. A new, better integrated, more forward-looking approach is needed, prioritizing projects that protect against future climate threats, enable our communities to be successful, and fairly and equitably protect lives and property for a safer future.

---

[1] Fact Sheet: The Bipartisan Infrastructure Deal, The White House. URL:https://www.whitehouse.gov/briefing-room/statements-releases/2021/11/06/fact-sheet-the-bipartisan-infrastructure-deal/. Accessed Nov 6, 2021.

[2] U.S. Infrastructure Grade, ASCE's 2021 Infrastructure Report Card URL: https://infrastructurereportcard.org/infrastructure-categories/. Accessed Nov 6, 2021.

[3] Fact Sheet: The Bipartisan Infrastructure Deal, The White House. URL: https://www.whitehouse.gov/briefing-room/statements-releases/2021/11/06/fact-sheet-the-bipartisan-infrastructure-deal/. Accessed Nov 6, 2021.

Government agencies charged with making the most of this generational investment and other investments in infrastructure are confronting large and challenging tasks.[4] Projects that may have been approved previously but are waiting for funding may need to be reevaluated to see if they are in line with new priorities. Decision-making agencies must also develop new frameworks to decide which projects to fund and which not to fund. Recognizing that agencies, states, and communities might benefit from a neutral convener, the National Academies of Sciences, Engineering, and Medicine, a science and evidence-based organization helping to address these challenges, decided to launch an initiative on infrastructure investment prioritization by convening a workshop on November 15, 16, and 18, 2021, aimed at trying to address these complex issues focused on infrastructure in the Gulf of Mexico region.

PROCEEDINGS ORGANIZATION

This proceedings is organized with, first, a description of the workshop format and processes in Part 1. Part 2 addresses the projects and the prioritization, and provides a description of the observations and key takeaways from each scenario-based discussion and from the capstone discussions, where the thinking was brought together. Part 3 focuses on the prioritization framework, especially the criteria that participants applied to the projects and how they think about how the pieces relate. Part 4 briefly describes participants' comments on the workshop, and on possible next steps to follow the workshop for both framework development and implementation. Appendix A presents the takeaways by domain. Appendix B lists all of the project ideas. Appendix C, written by Eleanore Douglas, describes the prioritization framework. Appendixes D, E, and F outline details about the workshop agenda, speakers, and participants, respectively.

Key takeaways from this workshop were cultivated at three levels of specificity: workshop-wide takeaways, scenario-specific takeaways and high-priority projects, and detailed takeaways and the comprehensive list of projects. To make it easier for the reader to find the takeaways, these categories are described in more detail below, along with guides to where they can be found in the proceedings.

1. **Workshop-wide Takeaways:** Workshop-wide takeaways include the nine key takeaways organized under three topics: (1) the necessity of community engagement, (2) the need to overcome long-standing obstacles, and (3) the need to move into the future. The topics and takeaways are mentioned in the Summary and expanded upon throughout the body of this proceedings. Another workshop-wide takeaway is the prioritization criteria that were developed and form the foundation of a prioritization framework for selecting projects that will increase infrastructure resilience in the Gulf region. They are located in the Summary and elaborated upon in Part 3. Appendix C includes an explanation of the background research and rationale for the approach to developing the criteria and corresponding framework.

---

[4] Although the largest allocations in the IIJA are for dedicated programs, like bridge replacements, the IIJA also has billions or tens of billions of dollars for more cross-sector or integrated programs that provide grants to improve resilience or prevent harm. Furthermore, the IIJA is only a piece of infrastructure spending by federal, state, and local governments and the private sector.

2. **Scenario-Specific Takeaways and High-Priority Projects:** Scenario-specific takeaways are trends and key ideas that emerged across multiple domains, specific to either the hurricane or protracted oil spill scenario. There are 23 scenario-specific takeaways, and they all map to the three topics listed above under Workshop-wide Takeaways. High-priority projects are those projects within each domain that either garnered the most votes across all macro-criteria or garnered an atypically high number of votes in a single macro-criterion. There are 46 high-priority projects. All of this information is located in Part 2.
3. **Detailed Takeaways and Comprehensive Project List:** Detailed takeaways are ideas that emerged in individual domains on individual days and even in individual exercises and were emphasized by workshop participants as being particularly important. There are 84 of these more granular takeaways, and they are listed in Appendix A. The complete list of all 306 project ideas that were identified during this workshop is in Appendix B.

All of the takeaways reflect comments and discussions at the workshop. Takeaways are presented as having been said by workshop participants. Readers should not regard these as carrying the weight of recommendations from the National Academies of Sciences, Engineering, and Medicine. The views cited are not necessarily consensus views of the group, and the group was not composed to meet National Academies' standards for study committees that make consensus findings and recommendations. The takeaways do reflect key ideas presented or discussed by one or more workshop participant(s), so the material presented here could be considered suggestions coming from informed individuals in the process.

CONVENING A WORKSHOP

To help prioritize among possible investments to improve the resilience of built infrastructure in the Gulf of Mexico region,[5] the National Academies convened a diverse group of experts for a 3-day interactive workshop in November 2021.

The purpose of the event, Investing in Resilient Infrastructure in the Gulf of Mexico Workshop, was to support the National Academies Gulf Research Program (GRP) in demonstrating and refining a process to help inform recommendations for prioritizing infrastructure investments across sectors and anchored in the Gulf region and its energy industry. The workshop had four main objectives:

1. Use two scenarios—a hurricane and a protracted oil spill—to start discussions on infrastructure vulnerabilities
2. Identify and prioritize projects or investments to reduce those vulnerabilities, thereby strengthening infrastructure resilience in the Gulf region
3. Identify criteria that would drive infrastructure investment decisions and prioritization

---

[5] There are many different definitions of infrastructure in use today. This workshop focused on the built environment, including human-fabricated physical structures and interventions in the natural environment. Social infrastructure is supported by these investments, and although it may require its own major investment, it was outside of the scope of this workshop.

4. Define criteria that will form the foundation of a repeatable prioritization framework for infrastructure investment decisions through cross-comparison of insights from each of the two scenarios

The workshop also introduced three policy questions to guide discussion:

1. What choices would maximize resilience return on federal investments in infrastructure and promote equity, fairness, economic effectiveness, and other goals and requirements?
2. How can we overcome obstacles to making the best investments in infrastructure?
3. How can we mitigate against private infrastructure assets becoming public liabilities?

Approximately 50 infrastructure, oil and gas, petrochemical, emergency management, and Gulf region experts from federal, state, and local governments; industry; non-governmental organizations; and academia participated in the 3-day workshop. Half of the participants joined Day 1 on November 15 for a set of in-person exercises focused on a hurricane scenario; half of the participants joined Day 2 on November 16 for a set of in-person exercises focused on a protracted oil spill scenario. The groups were split to ensure that they were small enough for every attendee to participate substantively and that appropriate expertise was represented in each group. Exercise materials included detailed scenario descriptions and chains of adverse events that could likely result from each scenario. Participants reviewed and updated that information, discussed vulnerabilities related to each scenario that impact local infrastructure, and then brainstormed and prioritized projects to address those vulnerabilities. They also discussed their considerations in making prioritization decisions. All participants joined Day 3 virtually on November 18 for a set of capstone exercises that compared and built upon ideas discussed in Days 1 and 2. See Figure 1-1 for an overview of this plan.

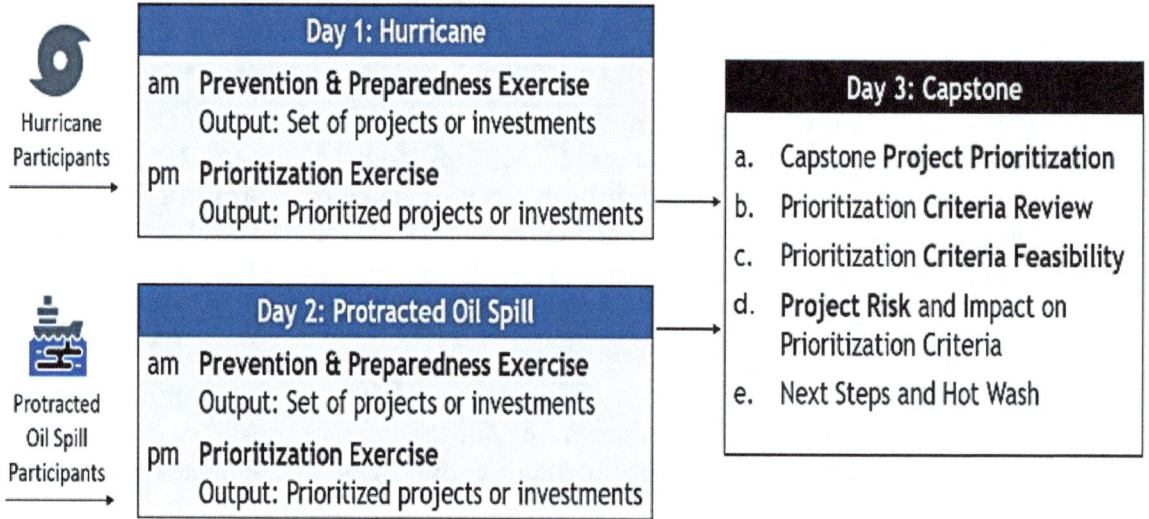

FIGURE 1-1 Overview of workshop plan.

Discussion and information collection was structured around four domains:

1. Oil, gas, and petrochemical (abbreviated as petrochemical) industry functions
2. Other infrastructure functions

3. Society's needs
4. Environmental protection

These domains, elaborated in Figure 1-2, provided a useful mechanism to organize inputs and feedback. On Days 1 and 2, participants joined small groups designed to include experts with diverse backgrounds, and all participants rotated through all domains. On Day 3, participants joined small groups composed of experts with similar backgrounds, and they spent their time in a domain that closely aligned with that background.

Petrochemical Industry Functions
- Production (Crude Oil)
- Refining
- Transport
- Retail

Society's Needs
- Housing
- Education
- Healthcare
- Commerce

Other Infrastructure Functions
- Communications
- Electricity
- Water
- Transportation

Environmental Protection
- Air
- Land
- Water

FIGURE 1-2 Domains around which workshop discussion and information collection was organized.

SETTING THE CONTEXT THROUGH KEYNOTES

The National Academies arranged for three keynote speakers to provide context for this event. Jason Tama, White House National Security Council Director for Resilience and Response, gave opening remarks during the first plenary session of the workshop. He opened the talk by noting that the timing of this workshop could not have been better; many factors that are fueling a "robust and certainly long-overdue dialogue on resilience." Specifically, Tama noted that we, as a nation, have been tested multiple times this past year, witnessing vulnerabilities in our infrastructure and being forced to reconsider what resilience is. These events, along with international and national-level developments such as the 2030 Strength and Resilience Commitment, adopted by the NATO allies at the 2021 Brussels summit,[6] as well as the signing of the Infrastructure Investment and Jobs Act, add momentum to—and provide resources for—efforts to build infrastructure resilience.

Tama pointed to the importance of spending the soon-to-be available IIJA funds "with resilience in mind" to build "safer and more secure and more resilient systems that Americans can rely on in times of crisis." When considering recovery, he encouraged participants to think of investment ideas that would help the nation "bounce forward" after incidents, so communities end up in a better place than they were previously. His perspectives on what can be learned from 2021 about resilience and what factors ought to be considered in infrastructure investment projects include the following:

---

[6] See https://www.nato.int/cps/en/natohq/official_texts_185340.htm?selectedLocale=en.

- **We have a shared responsibility to achieve collective resilience.** Tama pointed to the various supply chain shortages Americans have been grappling with since the beginning of the COVID-19 pandemic as a testament to the interconnected world we live in. The inability to produce or successfully source medical supplies, microprocessors, and fuel can impact critical infrastructure sectors and our national security, and be the difference between life and death. Only through collective resilience, which will require a "shared responsibility across all aspects of our society," will we be able to strengthen our infrastructure and thus, our communities.
- **Resilience should account for cascading impacts.** Tama shared a few examples of how our growing interconnectedness—be it through data, power, equipment, or people—has amplified cascading impacts during incidents. A ransomware attack on the Colonial Pipeline leading to fuel shortages in parts of the United States is a prime example of how a compromised IT system can have cascading impacts on other sectors of the economy. Power outages during Hurricane Ida impacted communications, oil and gas production, transportation, recovery operations, and more. Historical droughts and rising temperatures reduced our ability to generate hydropower and meet power demands, and they worsened wildfires. Tama encouraged participants to consider different approaches to resilience that will minimize cascading impacts within and across sectors.
- **Striving for equity in infrastructure investments is essential.** Tama spoke to the importance of working toward equity in infrastructure investment projects that will make disadvantaged communities more resilient, as they are disproportionately impacted during incidents and by the costs of recovery.
- **Establishing a common understanding of resilience can help drive action.** In concluding his remarks, Tama noted that "resilience is an easy buzzword word to throw around ... but we really need to be able to translate that to action." He asked the participants to develop a common definition for resilience and implementation principles that would help the government, private sector, and communities make actionable decisions to achieve the best possible outcomes.

Participants also heard remarks from Lt. General Thomas P. Bostick (U.S. Army, retired), Chairman of Bostick Global Strategies and a National Academies Steering Committee member. He opened by sharing insights about his time with the U.S. Army Corps of Engineers, where he advocated for a systems-level approach to infrastructure investment. He noted that resilience is not achieved by focusing on individual projects that treat parts of a larger system, but it is achieved by focusing on what is best for the system as a whole. He noted the importance of talking about resilience in exact terms to achieve actionable and measurable results. Without defining resilience and understanding the math, science, and engineering behind systems, the efforts to strengthen our infrastructure will be futile.

General Bostick also emphasized creating something that future generations would benefit from. He shared that only recently, in light of worsening natural disasters, has the conversation shifted from investing in protection systems to investing in resilient and risk reduction systems. He gave notes about what makes systems resilient, why resilience must be prioritized, and the importance of stakeholder engagement in decision-making. Some of these insights are summarized below:

Resilient systems should bend, but not break. Engineers design systems with a certain functionality to perform a desired goal. When planning for system risks in the past, the conventional wisdom was to not let risks cause a disruption to the system. While a worthy goal, General Bostick noted that it has proven to be counterproductive in light of recent incidents and the rising costs of recovery. By introducing resilience into the equation, engineers can make systems absorb the inevitable disruptions without breaking the system; systems should recover by adapting from the lessons learned and increase their functionality and be much improved.

- **There is a balancing act between protection systems and risk reduction systems.** General Bostick shared the example of how the Hurricane and Storm Damage Risk Reduction System (HSDRRS), which was built after Hurricane Katrina, helped reduce flooding during the 2011 flood compared with the 1927 flood that damaged the same region. With unlimited resources, engineers would have worked to prevent the flooding altogether. With limited resources, however, General Bostick noted that we will have to make pragmatic decisions on where we can protect and where it is better to reduce risk.
- **Stakeholder engagement and priority setting in infrastructure investment projects is crucial.** General Bostick shared his experience of working with outraged and battered communities after Superstorm Sandy, where key stakeholders were not involved or engaged in the prior infrastructure investment decision-making processes. Depending on the level of investments made, some communities fared better than others. Highlighting again the point about limited resources, General Bostick noted that "you're always going to need to set priorities and those priorities will leave people out.... They're not going to be any happier after the disaster, but if they're part of that decision to say, 'this is where we're going to spend our money,' then we're in a much, much better place."
- **He issued a call to broaden the concept of systems.** General Bostick concluded by encouraging the participants to broaden what we include in systems and to go beyond a specific sector or industry. Systems should account for "all the other impacts that keep people from living their day-to-day lives." He acknowledged that this would be a difficult undertaking because it requires many stakeholders, but by working together, we can create something "that will make all the difference in the days ahead."

On the final day of the workshop, Marcia McNutt, President of the National Academy of Sciences, delivered the opening remarks. She began by recognizing the passage of the bipartisan Infrastructure Investment and Jobs Act, which will invest more than one hundred billion dollars per year in infrastructure, giving our nation a rare opportunity "to improve the lives of our people for the coming decades and beyond." But the resources must be spent wisely and effectively, because even with these funds it will be impossible to execute every necessary or valuable infrastructure project. McNutt then shared her perspectives on some challenges to infrastructure investment that came up during Day 1 and Day 2 of the workshop, which are summarized below:

- **"Improving equity in outcomes is imperative."** McNutt acknowledged that communities of color have been disregarded in discussions of resilience for decades and suffer disproportionately more from hazards than other communities. Going forward, we must be inclusive and continue to invest in these communities, because "infrastructure is not a set-it-and-forget-it issue."

- **"We must bridge the gap between our vision for the future and needs of the present."** When designing systems, we need to account for future needs and emerging risks, and implement appropriate solutions now before the window of opportunity passes. As McNutt pointed out, "We cannot invest looking in the rear-view mirror." This will involve reducing red tape; harmonizing national, regional, and local interests; and overcoming structural and systemic challenges to how we currently approach infrastructure investments.
- **"We need to build back better to prevent legacies of failure in each natural disaster."** Infrastructure investments are front and center in everybody's mind after natural disasters have just occurred, and the easiest way to recover is by rebuilding what was destroyed. We must break out of this cycle by continuously investing in infrastructure when necessary and rebuilding in a way that allows systems to withstand shock as well as lessen harm to communities. Only by anticipating problems ahead of time and integrating lessons learned into our infrastructure investments can we build resilience.

THE EXERCISES

The National Academies sought to provide unique value from this workshop by maximizing discussion and interactivity. To accomplish this, the majority of time spent throughout the workshop was dedicated to a series of exercises in small groups organized around the four domains described earlier in this proceedings. As noted previously, on Days 1 and 2, the same set of exercises was conducted, and each focused on a different scenario: Day 1 focused on the hurricane scenario, and Day 2 focused on the protracted oil spill scenario. On Day 3, all participants focused on a set of capstone exercises that were scenario-agnostic, in order to compare and build on the takeaways from the previous two days.

The Prevention and Preparedness Exercise

The Prevention and Preparedness (P&P) Exercise was the first exercise held on Days 1 and 2. Participants used chains of adverse events that could likely occur in each scenario to prompt thinking and brainstorming of actions, projects, or investments that could prevent these adverse events from occurring in the first place or help the Gulf region prepare for and become more resilient to them or other events if they do occur. Workshop designers identified these chains of events as part of the scenario development prior to the workshop and displayed them on large poster boards. There were four boards for each scenario; each board represented one of the four domains used to organize workshop discussion and information collection. See Figure 1-3 for a sample board representing the Petrochemical Industry Functions domain for the hurricane scenario (Day 1).

Workshop designers divided participants into four small groups, intending to create groups of experts with diverse backgrounds. Each group reported to its first domain, where it was met by a facilitator and a scribe. The facilitator invited participants to first explore the entire board, reviewing the chains of events. Participants were invited to add adverse events to the chains of events preprinted on the boards, if they felt something important was missing. A few adverse events, impacts, and connections among them were added to the boards. Different participants gravitated to different areas of the board, which was expected and welcomed, typically driven by their areas of expertise and/or interest. Next, the facilitator asked participants to think about actions, projects, or investments that, if taken, would prevent these adverse events from happening in the first place or mitigate their impacts if they do occur. All projects were welcomed—as in some cases, a half-formed project idea may be better or more important than a fully mature idea—being "shovel-ready" is only one consideration. The readiness or maturity came up later when technical merit and feasibility were considered. Participants wrote down project ideas on self-adhesive notes and attached the notes directly to the board. The facilitator led discussion about the projects, attempting to elicit additional detail and inputs from other small-group members. After 20 minutes, the small groups rotated to the next domain and repeated the same activities. They also had the option to discuss and build upon projects suggested by previous groups. This exercise continued until all participants rotated through all domains.

This exercise was not conducted to provide a definitive or authoritative list of projects, and it did not seek consensus among the workshop participants. It allowed for the brainstorming of project ideas for further consideration and to be used in subsequent workshop exercises aimed at understanding what participants believe is important in making project prioritization decisions.

The Prioritization Exercise

The second and final exercise conducted on Days 1 and 2 was the Prioritization Exercise. Participants prioritized the projects brainstormed during the P&P Exercise using five prioritization macro-criteria: Environment, Economy, Society, Resilience,[7] and Project Governance. Workshop designers selected these macro-criteria after an analysis of other frameworks, best practices, and National Academies priorities, as detailed in Appendix C.

Participants remained in their small groups from the P&P Exercise, and they rotated through all four domains one more time. Upon arrival at a new domain, facilitators reviewed the boards, now populated with the projects that had been brainstormed by all groups earlier that day for the given domain.[8] Then the facilitator assigned each small-group participant a different prioritization macro-criterion and asked, "Which three projects provide the most value for, or have the greatest positive impact on, that criterion?" Participants cast their votes using different colored chips; they could assign more than one chip to the same project. (See Figure 1-4, a graphic that was available in hard copy to the participants during this exercise, outlining the five macro-criteria and corresponding chips, for more clarity about this exercise.) Then the facilitators prompted discussion, asking participants to explain their choices and most importantly, explain what factored into that decision. What aspects of that criterion were important?

---

[7] All of the criteria connect to resilience, and resilience is a key overall objective. The resilience criterion highlighted here was meant to prompt participants to rank the project according to its contribution to overall resilience, including adaptability and prevention of failures and exposures to physical harms. The criteria are elaborated in Figure 1–4 and Part 3.

[8] Prior to the Prioritization Exercise beginning, facilitators consolidated redundant projects, moved projects to more applicable domains when appropriate, and removed ideas that were not projects but rather observations.

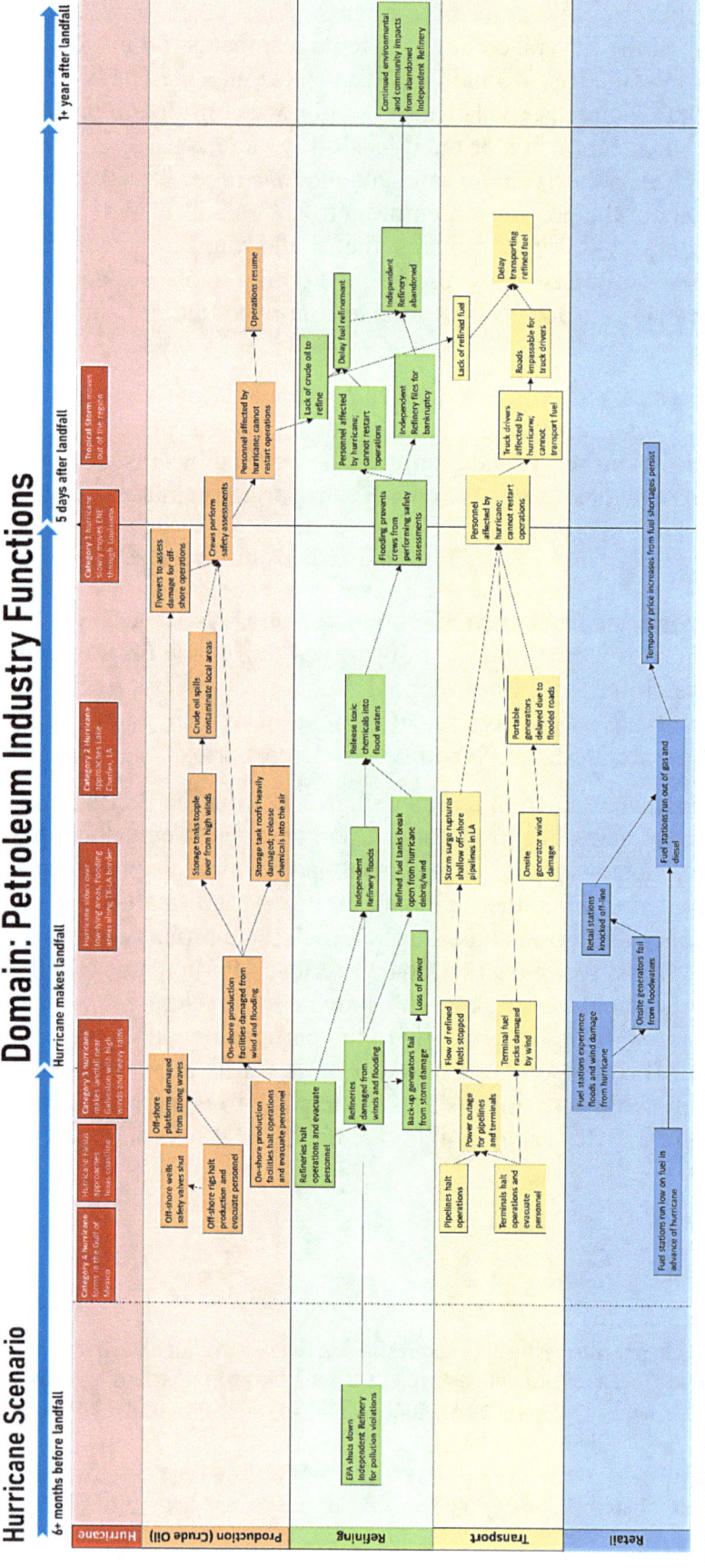

FIGURE 1-3 Sample poster board used to drive project brainstorming during exercises on Days 1 and 2 of the workshop.

The Prioritization Exercise

Which projects provide the most value for, or have the greatest positive impact on, that criterion?

 Environment
natural resource base, including land, water, and biodiversity in addition to the global conditions in which those resources reside, i.e. climate

 Economy
developed or developing human oriented activities, items and services, and the market in which they are bought, sold, or traded

 **Society**
qualities that adhere to individual human beings as individuals and communities: education, health, safety, diversity, equity

 Resilience
key qualities of resilience that infrastructure programs seek to imbue, e.g. robustness, redundancy, responsiveness.

 **Project Governance**
organization and management of groups and communities, to include leadership, oversight, transparency, bureaucratic dynamics, inclusivity re project development and implementation

 **Your Choice**
What else is important to you when prioritizing actions or investments that is not covered here?

FIGURE 1-4 Macro-criteria used in the Prioritization Exercise.

How did participants break the criterion down? Facilitators asked the participant assigned a given macro-criterion to lead off the discussion for that macro-criterion, and ultimately opened discussion to all participants on all criteria. Participants also shared when they felt other important prioritization criteria should be considered outside of the five predetermined macro-criteria. Finally, facilitators asked participants to weigh the relative importance of the five macro-criteria for the given domain when prioritizing potential projects. Participants assigned 100 percentage points across the macro-criteria, considering whether the macro-criteria should be weighed equally or whether some are more important than others, and if so, how much more important. Participants provided feedback using a handout provided.[9]

The Capstone Sessions

All participants joined together virtually on Day 3 for a series of capstone sessions and exercises to further build out and refine a set of criteria that would form the foundation of a prioritization framework for infrastructure investments. They compared outputs from both scenarios, elaborated on key criteria, and reexamined that key criteria in light of feasibility and project risk.

For Day 3, workshop designers organized participants into new groups; this time, intentionally grouping participants with similar backgrounds and expertise together and assigning the group to a domain that aligned to its expertise.

---

[9] The workshop designers and the National Academies recognize that there are shortcomings of simple linear weighting schemes. For example, they do not allow for contingent weighting (project B is only valuable if projects A and C are implemented at the same time). But the process and results are more about prompting the discussions and identifying issues than the actual scores.

The first session was the short Capstone Project Prioritization exercise. Workshop designers had analyzed the votes cast on Days 1 and 2 of the workshop, identifying approximately 12 high-priority projects per domain. Projects became part of this high-priority group either because they received the most votes across all macro-criteria or because they received an atypically high number of votes in one particular macro-criterion. During the first session of Day 3, each group prioritized the set of 12 projects for its assigned domain, using an online ranking tool that showed the compiled results of individual submissions in real time. Then facilitators led discussions to understand how participants went about making their decisions, priming them to think once again in terms of prioritization criteria, which would be the focus of the three subsequent sessions.

Next, participants moved into the longest session of the day, the Prioritization Criteria Review Exercise. Workshop designers had aggregated the sub-criteria—the elements of the macro-criteria identified by participants as important to their prioritization decision-making—during the Prioritization Exercise on Days 1 and 2, organized by domain and by macro-criteria. Approximately six sub-criteria per macro-criterion emerged as relevant to all domains; between zero and three additional sub-criteria emerged as key to each macro-criterion per individual domain. Facilitators presented participants with all sub-criteria for their domain (see Figure 1-5 for an example), and participants identified the five sub-criteria that they believed were most important in each macro-criterion category. Then they engaged in a detailed discussion to elaborate on how they defined each sub-criterion for the purpose of prioritizing projects that would increase infrastructure resilience in the Gulf, and why they felt the sub-criteria were especially important.

FIGURE 1-5 Sample presentation of five macro-criteria with corresponding sub-criteria for the Other Infrastructure Functions domain, for further review and discussion.

The Prioritization Criteria Feasibility session was the third session of Day 3. This was largely an open discussion about the feasibility of incorporating various prioritization criteria into a prioritization framework. Facilitators asked participants to think through impediments to feasibly applying these sub-criteria during the prioritization processes—for example, impediments resulting from legal, financial, time, and data constraints.

The capstone sessions ended with a session entitled Project Risk and its Impact on Prioritization Criteria, a brief discussion about project risk and what implications that has, if any, on the prioritization criteria discussed so far. Facilitators selected three high-priority projects from the morning's Capstone Project Prioritization Exercise that were clear and covered a range of topic areas. They asked participants to brainstorm risks to the projects' success and mitigation strategies to address those risks. Finally, the groups discussed whether any of those risks suggested new or modified criteria that should be included in the prioritization framework.

THE SCENARIOS

Workshop designers prepared detailed scenario narratives to help prompt project brainstorming and prioritization considerations by participants during the workshop. They provided these narratives, along with executive summaries capturing key highlights of each narrative, to participants before the workshop as read-ahead materials. The following are the executive summaries that were provided, offering highlights about these notional events.

A Retrospective: Texas and Louisiana Still Recovering from Category 4 Hurricane Falsus One Year Later

Hurricane Falsus, the first major hurricane to strike Texas since Hurricane Harvey in 2017, made landfall on August 17, 2021 (see figure below). As the hurricane approached land, the National Hurricane Center forecasted that Falsus would stall over Southeast Texas and Louisiana, causing widespread flooding and devastation to infrastructure. These direct impacts soon cascaded to affect the environment and needs of area residents. We summarize the major impacts within the following four domains:

Petrochemical Industry Functions — The industry prepared for Hurricane Falsus by shutting down offshore and onshore platforms, evacuating oil rig workers, and closing refineries. The pipelines and terminals in the area halted operations. While offshore infrastructure damage was limited, Hurricane Falsus' impacts to onshore petroleum infrastructure were catastrophic. Refineries flooded, lost power, released toxic chemicals into floodwaters, and caused localized oil spills. Terminals, pipelines, and retail fuel stations lost power and sustained damage, and many remained inoperable long after the storm.

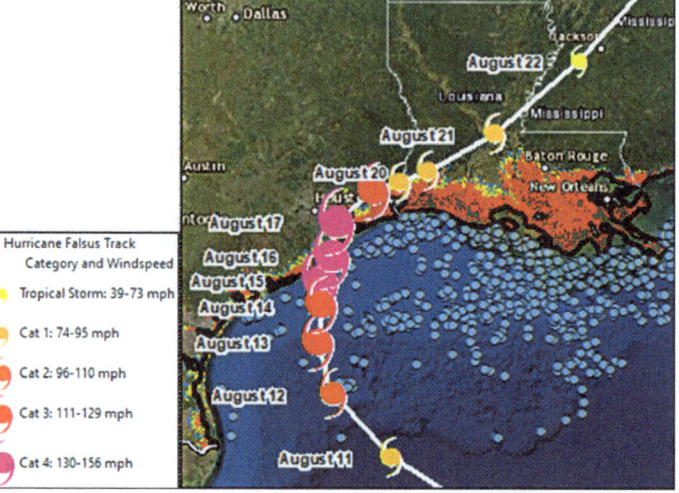

Track of Hurricane Falsus

Other Infrastructure Functions — The 130-mile-per-hour winds damaged electrical transmission lines, substations, and switching stations, leaving residents without power long after the hurricane. Strong winds disabled cell towers and internet services, leaving many residents unable to reach emergency dispatchers. Water and wastewater treatment plants were flooded and released untreated wastewater, leading to widespread contamination of surface water. Highways and local roads were closed because of debris, downed telephone poles, flooding, subsidence, and damaged bridges.

Society's Needs — Flooding and loss of power, running water, and sewer disrupted basic infrastructure functions. Chemicals in floodwater and mold made homes uninhabitable, forcing out families. COVID-19 outbreaks occurred in emergency shelters. Families with modest means or insufficient insurance could not afford home repairs or mold remediation. Hospitals offered delayed or degraded care because of flooding, power outages, and staff shortages. Schools closed for weeks.

Environmental Protection — Storm surges caused extensive land loss in Louisiana, which worsens with each successive storm. Local oil spills, flooded industrial areas, and Superfund sites released hazardous materials onto farmlands and into bayous, contaminating crops and causing animal die-offs. Residents in flooded areas tell mayors that their well water is still not safe to drink. A damaged and bankrupt oil refinery sits idle, a looming threat when the next major storm hits the area.

Submarine Landslide Triggered Oil Spill Near Port of Galveston; Recovery Ongoing

Acme Energy, LLC, is a privately held oil and gas company headquartered in Houston, Texas. Acme Energy owns Platform A, which is located about 20 nautical miles southeast of the entrance to Galveston Bay. The platform has 28 oil and gas wells extending from the structure into the seafloor. In March of 2012, Acme Energy halted production at Platform A and shut all active wells while it assessed the future viability of production at this location. In the summer of 2012, the seafloor beneath Platform A destabilized, causing an underwater regional slope failure. The platform fell to the seafloor, damaging the jacket structure beneath it. The platform, jacket, and tangle of pipelines and 28 wells were buried and/or damaged.

Several days later, a commercial fishing vessel spotted a large oil slick (figure at right), noted the absence of Platform A, and reported the spill to the US Coast Guard (USCG). The USCG established a Unified Command and began to direct mitigation actions. News outlets immediately began reporting on the potential for environmental damage and on similarities to other recent oil spills, such as Taylor and *Deepwater Horizon*.

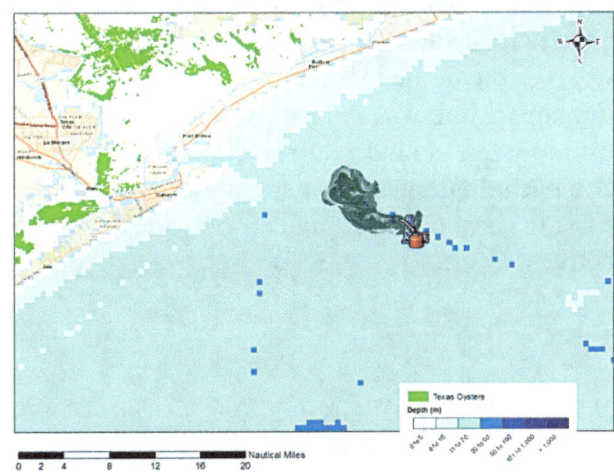

Oil sheens visible at the site of Platform A, moving toward Port of Galveston

One year later, Acme Energy completed installation of a subsea containment system by placing domes over three leaking well sites and began efforts to plug the leaking wells. Although those efforts were successful in reducing the amount of oil leaking to the surface over time, oil continued to leak and cause visible sheens. Prevailing winds pushed oil toward the Port of Galveston, its beaches, and sensitive fisheries along the coast and in the bay. Environmental groups conducted their own monitoring efforts to ensure accountability. News reports kept the spill front and center in people's minds.

Three years later, Acme had plugged many of the leaking wells. Still, oil continued to leak to the surface causing visible sheens. Despite no evidence that oil was reaching shorelines and causing significant damage to the environment at the time, the constant media coverage took a toll on the tourism industry in the Galveston area. Fishing charters reported that business had declined by 40 percent compared with pre-spill levels and beach communities similarly reported declines in business. Similar impacts were evident on commercial fishing.

Five years later, it became clear that there were problems with the original containment system, as oil sheens were still routinely reported at the site and the USCG ordered Acme Energy to design and install a new system. Impact to the economy in Galveston, Texas, was also evident. Some tour boat and charter boat operators went out of business and there were closures of restaurants, shops, and hotels in beach communities. Commercial fishing companies reported decreases in demand, causing prices for Gulf shrimp and shellfish to decline. Cruise ship departures declined and many cruise lines have moved their operations.

The spill and its ongoing impacts to the Gulf region continue today.

THE HYBRID FORMAT

The hybrid format used for this workshop—which mixed in-person and virtual sessions—successfully enabled workshop designers to collect substantial and high-value outputs, and the 1-day gap between the second and third days of the workshop allowed sufficient time for analysis and distillation of Day 1 and Day 2 outputs to set up for Day 3. Participants provided positive feedback on the hybrid format during the hotwash at the conclusion of Day 3, which was a session focused on evaluating what went well and what should be adjusted if this type of workshop were held again. By holding Days 1 and 2 in person, participants had an opportunity to forge new relationships and take advantage of the energy created by the dynamic interaction of a group of dedicated experts brought together by a common goal. By holding Day 3 virtually, participants' time was respected, allowing them to travel back to their homes. It also accommodated the individuals who, due to either personal or professional conflicts, were not able to participate in an in-person event. By utilizing online collaboration tools, Day 3 discussions and inputs remained rich and insightful, ultimately achieving all of the workshop objectives.

Part 2

Project Identification and Prioritization

As described in Part 1 of this proceedings, Days 1 and 2 of the workshop used a series of in-person exercises to guide participants into identifying and prioritizing project ideas that would likely increase infrastructure resilience in the Gulf of Mexico region. Project brainstorming and prioritization was organized by scenario and by domain. This section offers observations and key takeaways that emerged during those exercise discussions. These are ideas that came up repeatedly across different small groups and more than one domain. These observations and key takeaways align with and elaborate on the key takeaways outlined in the Summary. They are summarized in Figure 2-1 and described in greater detail below the figure. The workshop participants are experts, but they were not asked to research, suggest, and vet detailed project ideas, so those ideas should not be taken as ready-to-use investments. As was stated throughout the workshop, the project ideas themselves were not as important to the value of the workshop as the role they served as a tool to show the processes by which participants conceived and weighed considerations for investment prioritization. More specific takeaways broken down by individual domain and individual exercise can be found in Appendix A. This section concludes by listing the projects that were prioritized highest for each domain, either because they garnered the most votes across all five macro-criteria or because they garnered an atypically high number of votes in a single macro-criterion. The full list of all 306 projects ideas identified during the workshop can be found in Appendix B.

Necessity of community engagement	
Emphasizing local leadership **Promoting community involvement** **Improving communications and transparency**	
— Engage community early in decision-making for better decisions and stakeholder buy-in — Education campaigns are needed for the community and in schools	— Engage community early in decision-making for better decisions and stakeholder buy-in — Effective public information campaigns are needed to combat mis- and disinformation and to build and maintain public trust — Enhance the relationship between the government and the media
Call to overcome long-standing obstacles	
Moving from survivability to thrivability **Focusing on prevention in addition to response** **Incorporating resilience and equity consistently**	
— Address basic survival, which is still at risk in the Gulf region and is a fundamental priority — Move toward more reliable broadband access — Improve workers' safety and security — Focus on prevention. Stop putting infrastructure in low-lying areas — Make both short-term and long-term infrastructure investments — Conduct more preplanning	— Improve or expand the Oil Spill Liability Trust Fund for cleanup — Focus on prevention. Remove aging and abandoned infrastructure to reduce risk — Conduct more coordination across oil, gas, and petrochemical industry stakeholders, the government, and the public, up front
Need to move into the future	
Building up data and analysis **Embracing technology and modernization** **Balancing regulations versus incentives carefully**	
— Adopt wider and more varied use of sensors — Invest in next-generation monitoring — Build up data and analysis to inform better decision-making — Future regulations and incentives need careful consideration and constant monitoring	— Improve monitoring capabilities — Invest in R&D to continuously improve technology — Build up baseline data — Bolster data and information collected through forward-looking research — Smarter regulation of the petroleum industry is key to a resilient Gulf

FIGURE 2-1 Overview of workshop-wide and scenario-specific key takeaways stated by some participants.

OBSERVATIONS AND KEY TAKEAWAYS: HURRICANE SCENARIO

Necessity of Community Engagement

Engage Community Early in Decision-Making for Better Decisions and Stakeholder Buy-In

Understanding who the affected populations are and what their needs are is vital to building community and infrastructure resilience to hurricanes. Therefore, participants said, local leadership in decision-making, including response operations, must be increased. Health and housing projects in particular need to be sensitive to and prioritize the needs of traditionally underserved, highly vulnerable populations. Local leaders who can represent these needs must be involved in related decision-making. For any projects to be successful, project teams need buy-in and input from the local communities, who often hear of grand projects that are not implemented or who typically have seen the negative aspects of projects that have been implemented around them. Early and frequent engagement is critical to share information and elicit community inputs to project-related decision-making.

Education Campaigns are Needed for the Community and in Schools

Education is very important and often overlooked because physical-built projects get more attention, and the human side can be lost. Residents need to be made aware when new services or amenities are developed and how to use and access them, workshop participants said. The community needs more and updated education campaigns about what impacts to expect from disasters that may take place, what backup systems are in place to address them, and what actions they should take under various conditions. Communities need more education on flood maps and how that information applies to homes that may flood or roadways that are likely to be inaccessible. Communities and policy makers would benefit from education about what resilience is, why it is important, and how to achieve it. Some participants suggested development and incorporation of a disaster science curriculum for schools, possibly including disaster recovery training for youth.

The Call to Overcome Long-Standing Obstacles

Address basic survival, which is still at risk in the Gulf region and is a fundamental priority. Workshop participants said that priority must be placed on investments that restore basic needs to the population post-storm, emphasizing first that every individual has access to clean drinking water. Next, recovery efforts need to establish functioning wastewater systems, communications systems, and safe housing, meaning a return to pre-hurricane habitability (but see numerous examples in subsequent sections of how to improve the current conditions, rather than maintaining them). The power grid must get back up and running immediately to provide a return to stability and sense of normalcy, and access to fuel at retail stations must be restored. Even better, investments should focus on preventing these issues from happening in the first place.

Move Toward More Reliable Broadband Access

Ideally move to 5G networks, throughout the region, focusing on coastal communities that are repeatedly impacted by storms, as well as areas with high social vulnerability. Workshop participants said that without reliable communications, access to critical post-storm functions, such as remote learning and online commerce, is compromised. People must be able to communicate with their close friends and family post-storm to promote a sense of calm. Investments should encourage and expedite this progress.

Improve Workers' Safety and Security

Workshop participants said that improved technology, addressed further under The Need to Move into the Future section, is a major contributing factor to protecting workers. Using better sensors, we can get information to workers sooner regarding which roadways are flooded so that they can get to their jobs more safely. Adoption of "smart fuel tanks" will signal the need for repairs earlier, leading to less dangerous working conditions. Increasing the use of remote technologies to gather data and monitor for damage and debris will keep workers from having to enter high-risk zones to collect such information themselves. In addition to benefits for workers from technological improvements, putting more emphasis on getting people back into their homes post-disaster will support a faster, safer return to work by the affected population. It is even better if housing can be made more resilient through stronger building codes and building materials. Finally, providing special identification cards for workers who have to support a disaster response will facilitate their ability to do their jobs, facilitating their access to job sites within the impacted zones.

Focus on Prevention

Stop putting infrastructure in low-lying areas. Areas that storms will impact most frequently and severely are known, so people should stop building electricity and water infrastructure, chemical plants, and even oil tanks and refineries in these areas, workshop participants said. The capability to build elsewhere exists. Even if the initial investment is substantial, it is a worthwhile investment when considering the toll on society when these sites are damaged and/or inoperable. Also, much more research is needed into the future impacts of climate change and when new infrastructure is being designed and built, and that research should be taken into account, including where future impacts will occur.

Make Both Short-Term and Long-Term Infrastructure Investments that Improve upon Previous Conditions

Shorter-term projects with immediate results—usually related to hurricane response—are important even if the benefits to individuals are only marginal, as longer-term projects can take so long to have visible impact. Right now, communities need to see concrete successes, participants said. However, focus on the long term must be the priority. Projects should have positive long-term impacts on the well-being of communities and ecosystems. They should focus on prevention or avoidance of downstream harm and ultimately reduce the need for additional, response-oriented projects. Both short- and long-term projects must ensure, however, they are improving upon the conditions that caused the infrastructure to fail in the first place. They must create conditions that are healthier, safer, and more equitable, to drive resilience.

Conduct More Preplanning

There will always be a shortage of critical supplies during a response; therefore, it is critical to decide beforehand who gets priority access to them and to do so equitably. This includes preplanning and communicating local and national priorities, so that national requirements for recovery, which local leaders may not even be aware of, do not derail response operations once they are underway. The latter has occurred, for example, when distributing fuel in the aftermath of past disasters. Pre-event prioritization of projects is also necessary to ensure prudent allocation of limited resources for investment after an emergency.

The Need to Move into the Future

Too many projects and programs are stuck in the past, focused on fixing or restoring outdated infrastructure designed for environments that have changed. These takeaways focus on enabling communities to meet future needs.

Adopt Wider and More Varied Use of Remote Sensors Across the Board

Adopt remote sensors, especially in underserved and underprivileged areas. Sensors increase situational awareness and provide more data to make better, faster decisions; they help protect workers; and they enable a faster, more efficient and effective response to adverse conditions. Remote sensors will be exposed to disaster conditions and many will get destroyed during storms, but that is understood as the price of doing business; it should be factored in during installation so that enough remain to get the job done, participants said. After Hurricane Ida, remote sensors provided a robust hazards warning system and enabled the reconstruction of events. Examples of sensors that are priorities to be used in greater numbers include, but are not limited to, the following:

- Flood sensor systems on roadways to indicate flooding and flash flooding
- Sensors on fuel tanks, or smart tanks, to indicate earlier when repairs are needed
- Remote sensors that enable measurement of disaster-related conditions such as water levels, wind speed, and debris, installed on electricity and communications infrastructure at key locations of interest
- Networks of staff gauges, which measure water levels, to augment wireless sensors and enable large-scale flood reporting
- Sensors on gas-powered electricity generators that monitor carbon monoxide emissions
- Sensor networks that monitor pipelines and spills, indicating, for example, near-surface pollutants at estuaries

Invest in Next-Generation Monitoring

This includes such things as remote sensing and autonomous technologies, in addition to the sensors described above, and also increased and better use of drones, LIDAR (Light Detection and Ranging), radar, and so forth. During Hurricane Harvey, for example, several areas with chemical spills were inaccessible, so samples could not be taken because of the human health risks. Autonomous technologies could fill these gaps.

Build Up Data and Analysis to Inform Better Decision-Making

Many datasets that can benefit preparedness, mitigation, and/or response exist in silos, are poorly socialized, or simply are not leveraged in decision-making or in communications with the public, workshop participants said. This problem is exacerbated by the numerous jurisdictions (and authorities) involved in the Gulf region. Baseline data for detecting changes and impacts caused by hurricanes, sea level rise, and other hazards are needed. Participants said that key infrastructure system data should be mapped out, consolidated into a central repository, and made accessible to those who need it. And the nation must increase its understanding of the damage oil, gas, and petrochemical incidents cause.

Future Regulations and Incentives Need Careful Consideration and Constant Monitoring

Overreliance on the private sector to address adverse impacts related to a hurricane without incentivizing actions from the government may be misguided, participants said. Prudent use of both regulations and incentives—"sticks and carrots"—could be good for both industry and the public sector. Incentivization of private efforts can be precarious, though; government provision to private partners can result in modest or disappointing payoff. To avoid this pitfall, it is important to continually assess projects' effectiveness and hold partners accountable; otherwise, this can become subsidizing nonlocal business with little or no benefit to the community.

OBSERVATIONS AND KEY TAKEAWAYS: PROTRACTED OIL SPILL SCENARIO

Necessity of Community Engagement

Engage Community Early in Decision-Making for Better Decisions and Stakeholder Buy-In

Workshop participants said that to be effective, projects must be led locally whenever possible, and anyone impacted by the project must be informed about it and invited to comment on it. Public understanding of project implementation must be encouraged by project teams explaining what is happening and how money is being spent. Research must be conducted to better understand the needs of socially vulnerable or traditionally underserved communities during and after disasters. If public services have to be cut after a disaster due to budget constraints, those cuts should be based on community input.

Effective Public Information Campaigns Are Needed to Combat Mis- and Disinformation and to Build Public Trust

The public must understand what is happening during a spill and its cleanup, including how to prepare for one, what actions to take if it does occur, and what happens to oil that has to be cleaned up, including its impacts on the environment. Workshop participants said that campaigns should focus on the local population, but also tourists and visitors. The campaign should encourage revenue-generating activities that contribute to community and business resilience in the wake of a spill, such as tourism and seafood consumption, if safe to do so. To build and maintain trust with the public, all levels of government must communicate faster and with greater detail and clarity on these issues.

Enhance the Relationship Between the Government and the Media

Workshop participants said that the government and its partners must enhance their relationships with the media and share more information. Coordinated information campaigns should be developed, including the development of trusted information sources and identification of trusted community messengers. Progress on cleanup efforts should be shared through social media and other channels, demonstrating to the public that concrete actions have been taken and progress is being made.

The Call to Overcome Long-standing Obstacles

Improve or Expand the Oil Spill Liability Trust Fund for Cleanup

Conditions for drawing on the Oil Spill Liability Trust Fund should be revisited, updated, and set in advance, including identifying practical triggers to access the funds, workshop participants said. It should be administered by a third party, such as a committee, nonprofit, or academic institute, through a trust. Its management must be transparent and nimble. Further, this fund can run out during a catastrophic spill, so a comprehensive program with alternative funding sources must be established as well, such as grants, revolving funds, and low-interest loans—it should be a diverse portfolio. Finally, the improved fund is vital, but it should be treated as a stopgap measure that is used if the responsible party goes bankrupt and cannot meet all obligations.

Focus on Prevention

Remove aging and abandoned infrastructure, such as the 18,000 miles of unused pipelines, to reduce risk and prevent future environmental degradation and pollution. Also, areas vulnerable to submarine landslides or other accidents are known and should not be auctioned off for offshore drilling. Additionally, the Bureau of Ocean Energy Management's leasing standards and requirements should include provisions that would ban the abandonment of offshore infrastructure and make the organization responsible for its assets' maintenance from cradle to grave, workshop participants said.

Conduct More Coordination Up Front

Given the multiple stakeholders involved in oil spill detection and response, prioritize projects that improve coordination, collaboration, and information sharing among the relevant parties and with the public. Hold more exercises with state or local officials in charge. Involve industry early in the process to promote transparency and trust. Increase training with all the relevant stakeholders about oil spills. Create a dedicated budget for these coordination efforts, workshop participants said.

The Need to Move into the Future

Improve Monitoring Capabilities

Workshop participants said that the nation needs to develop new methods, probably through sensor technology, to monitor the network of pipelines. Deploy sensors to detect oil and to report on chemical composition of contaminants to inform response and recovery efforts. Monitor aging and abandoned infrastructure. Add sensors to pipelines to detect microfractures, and to speed repair efforts.

Invest in R&D to Continuously Improve Technology

R&D investments must be made in technologies that would prevent and better contain spills, workshop participants said. This includes new oil spill barrier designs; new methods for skimming; product design for environmentally friendly surfactants and dissolvents; methods for burying flow lines, pipelines, and fiberoptic cables; and improved rock dumping or similar protective actions for flow lines.

Build Up Baseline Data

A much better understanding of baseline data in the Gulf of Mexico, including health and economic data particularly among socially vulnerable populations, is needed to understand the impacts of a spill and to track and monitor the progress of projects designed to address them. Workshop participants said that studies should focus on current or natural environmental conditions, which are lacking and prevent assessment of an oil spill's impact on the environment and how protracted oil spills might or might not affect the environment. Specific attention should be paid to direct impacts from protracted oil spills to beaches and aquatic environments near shore, marine ecosystem impacts of slow-release oil spills in areas that already have natural oil seepage, and impacts on health and human safety from cleanup operations. Community members and local organizations should be empowered to use techniques such as crowdsourcing and geotagging to support data collection.

Bolster Data and Information Collected Through Forward-Looking Research

Workshop participants said that the research should include long-term epidemiology studies to understand impacts on human health; more research into modeling seismic activity to predict where landslides may occur; assessments of the potential impacts of landslides on wells to determine which are most at risk and need additional reinforcement to prevent leaks; and an ability to determine, with greater accuracy and more authority, information about a spill, such as the volume and impact of the spill, in ways that cannot be challenged by competing interests.

Smarter Regulation of the Petroleum Industry with Better Enforcement Is Key to a Resilient Gulf

Workshop participants said that attention and investments should support and implement smarter regulation, from stricter permitting—which should include requirements for recovery plans, better monitoring, and funds set aside for cleanup—to stronger government enforcement authorities and expertise to compel the responsible party to clean up in a timely manner. Oversight of legacy and abandoned infrastructure is another key item to address through stricter permitting. The cost of maintaining and servicing aging petroleum infrastructure is a real challenge and a reason some owners walk away from it, resulting in ongoing public liabilities. Corporations must be held responsible for environmental degradation and societal harm resulting from accidents of all scales. The Bureau of Safety and Environmental Enforcement needs more enforcement authority. The U.S. Coast Guard has more authority, but does not have oversight of subsea infrastructure. Legal changes are necessary to allow for better enforcement of regulations.

HIGH-PRIORITY PROJECT AND INVESTMENT IDEAS

As discussed previously, the Prevention and Preparedness Exercise conducted on Days 1 and 2 of the workshop focused on brainstorming project ideas that would increase infrastructure resilience in the Gulf region. The Prioritization Exercise focused on prioritizing those projects. Table 2-1 provides the set of project ideas for each domain that were prioritized highest across both scenarios by the participants. This means they either received the most votes across all five macro-criteria—Environment, Economy, Society, Resilience, and Project Governance—or they received an atypically high number of votes in a single macro-criterion. Note that these are sample project ideas identified by this one particular group of expert stakeholders. As noted previously, the workshop participants were not asked to research, suggest, and vet detailed project ideas, so those ideas should not be taken as ready-to-use investments, nor should they be construed as a definitive list of projects to be implemented, but rather as a starting point for further thought. The complete list of all 306 project ideas brainstormed during this workshop can be found in Appendix B.

TABLE 2-1 High-Priority Projects across Both the Hurricane and Protracted Oil Spill Scenarios

Domain	High-Priority Project Ideas
Petrochemical Industry Functions	• Increase local leadership in emergency response • Advance coordination of national power restoration priorities • Require and incentivize backup power for high-priority fuel stations • Bury power lines • Pass regulations establishing environmental, social, and governance (ESG) mandates for corporations • Invest in predictive modeling and forecast movement of oil and dispersants • Expand and/or refine the mandated cleanup fund paid for by the petroleum industry • Develop better, faster public information-sharing campaigns to improve public trust • Implement stricter industry owner permitting requirements • Treat fuel stations as "anchor" institutions, prioritizing them for backup power • Use flood sensor networks to map flooded areas
Other Infrastructure Functions	• Harden, bury, or elevate electrical infrastructure • Protect, harden, and/or elevate water infrastructure • Transition to 5G with hardened towers and battery backups • Develop a public education/emergency preparedness campaign, including likely impacts to infrastructure and what actions individual should take, when, and how • Invest in R&D for new oil spill barrier construction • Study impacts of climate change on infrastructure • Identify and map infrastructure systems, and make this information available in a GIS (geographic information system) tool

Domain	High-Priority Project Ideas
	• Conduct a community-wide needs assessment before cutting services in the wake of declining local budgets • Expand broadband access, start with coastal areas • Develop a data center for citizen and community data • Develop a Strategic Water Reserve
Society's Needs	• Develop plans for the communication of science, including messaging to and awareness for the public • Rebuild and repair transportation infrastructure anticipating future floods • Develop incentives for green technology and disaster-resistant building materials • Conduct baseline studies for community health and needs, especially in underserved communities • Reevaluate and streamline Federal Emergency Management Agency (FEMA) processes in general, for example, housing, flood insurance, incentives • Fund and overhaul FEMA flood maps for long-term planning in a changing environment • Conduct research on alternative oil response technologies to prevent and/or mitigate adverse impacts • Regrade roads with more adaptive materials, for example, porous materials, living streets • Incorporate community engagement into planning for communities and schools • Develop and implement training and education programs to improve economic mobility • Conduct long-term studies on the impacts of oil spills, including comparisons of protracted oil spills versus those similar to Deep Water Horizon • Create and expand programs to support local industries after disasters
Environmental Protection	• Build and protect natural protection such as coastal systems, reefs, wetlands, and sea grasses • Design and implement restoration projects for wetlands and disturbed habitats • Conduct epidemiological health studies to assess chronic impacts of pollutants on coastal communities • Change construction standards beyond the 100-year flood zone • Develop projects to improve the health of ecosystems to make more resilient coasts and barriers • Increase applications of remote sensing to monitor the impacts on air, land, and water • Improve geospatial data use and sharing for mitigation and early response • Change Bureau of Ocean Energy Management leasing to include cradle-to-grave infrastructure and account for equity • Invest in R&D product development of environmentally friendly biodegradable dissolvents and surfactants

Domain	High-Priority Project Ideas
	- Create and enhance protection systems for wastewater plants against flooding
	- Reassess the fisheries management process to allow sufficient time for recovery
	- Increase community resilience by providing mental and physical health support services

Part 3

Prioritization Framework

As described in Parts 1 and 2 of this proceedings, Days 1 and 2 of the workshop used a series of in-person exercises to guide participants into identifying and prioritizing project ideas that would likely increase infrastructure resilience in the Gulf of Mexico region. These brainstorming sessions were scenario specific and included facilitated discussions designed to elicit inputs from workshop participants on how they defined and distilled five macro-criteria—Environment, Economy, Society, Resilience,[1] and Project Governance—into sub-criteria, for the purpose of prioritizing infrastructure resilience projects. Participants also considered the relative importance of the macro-criteria. Workshop designers analyzed inputs collected on Days 1 and 2, identified the sub-criteria that emerged through these discussions (both across domains and specific to individual domains), and presented this information to four small groups on Day 3 for in-depth discussions about each sub-criterion. These discussions included whether each sub-criterion should be included in a final set of criteria that would form the foundation of a prioritization framework for infrastructure projects in the Gulf region, how each sub-criterion should be defined, how feasible it is to assess projects against these sub-criteria in an expedient manner, and whether explicit consideration of risks for project implementation introduces any additional sub-criteria.

This part includes the master set of criteria that would form the foundation of a prioritization framework, suggestions for the relative weighting of the macro-criteria, and details on how each individual sub-criterion might be defined based on open-source literature and participant comments. Appendix C describes the research and rationale that drove the early development of this approach and how the master set of criteria fits into that approach.

THE PRIORITIZATION CRITERIA

Figure 3-1 depicts the master set of criteria, an aggregation of the criteria reviewed by each of the four groups on Day 3 of the workshop that supports a prioritization framework for infrastructure resilience projects in the Gulf region. More detailed explanations of the sub-criteria are provided later in this section. Participants' perspectives on which criteria were more and less important than others varied, but there was only limited questioning about whether they should all be included in some form. Several participants suggested that technical merit or feasibility of a project should simply be a prerequisite of any project proposal and not assessed specifically as a part of Project Governance; others discussed if Resilience should be included as a stand-alone criterion. Overall, however, the general takeaway from Day 3 was that this set of

---

[1] Resilience as a concept is part of all of the other criteria and is a key overall objective. The resilience criterion concerns adaptability and prevention of failures and exposures to physical harms.

criteria in its entirety is worth inclusion in a prioritization framework and worthy of further attention.

When discussing feasibility of applying the criteria to project prioritization, participants offered several recommendations and reminders. They promoted leveraging existing datasets whenever possible and considering a two-level approach to the assessment of the sub-criteria. Level 1 is a binary assessment: Does the project address and satisfy the sub-criterion or not? Level 2 is open-ended: How does the project meet the sub-criterion and to what extent? At the same time, however, participants noted the burden that an overly elaborate prioritization framework would impose on project proposers, many of whom may not have an abundance of resources available to them to support the development of complex proposals, thereby exacerbating concerns about issues of equity. Workshop designers concluded that careful attention should be paid to asking what is truly necessary for effective assessment. Finally, the session on Day 3 dedicated to the identification of project risk and how that risk might impact the criteria, revealed the need to add a single new sub-criterion, "accounting for cyber/physical security implications."

Environment	Economy	Society	Resilience	Project Governance
• Advancement of scientific understanding of the environment • **Biodiversity** • **Climate change** • Degradation prevention • **Do no harm / reduce harm to environment** • Land use • Maintenance • Mitigation • Persistence of efforts • **Riverine and coastal erosion** • Safe living and working spaces • Sustainability • Understanding of environment baselines and changes • **Water/air quality**	• Business continuity • **Commerce support** • Continuity/recovery of economic activity • Cost-effectiveness • **Economic development** • **Employment** • Equity • **Financial need** • **Individual support** • Labor rights and working conditions • Multiple impacts • Proximity to major economic and/or residential centers • Spending drivers • **Tax base**	• **Access to information / awareness** • Benefits go to underserved communities • **Community engagement** • Connections • Cultural value • **Education** • **Equity and inclusion** • **Health and welfare** • **Public safety/order** • Public trust • Return to stable society/normalcy • Social and political acceptability • Survivability (e.g., water, basic needs)	• **Adaptability** • Expedited recovery • **Exposure reduction** • Focus on preparedness and response • **Hardening of infrastructure** • Info management and comms • **Interdependence with other systems** • Long-term impact • Planning and prep (prevention focus) • Rapid deployment for response • **Redundancy** • **Supportive of the system / network** • Understanding AND action	• Alignment with existing plan • Accounting for cyber/physical security implications • **Community inclusion** • **Data and information sharing** • **Feasibility** • Measurable • **Outcomes focus—effectiveness, return on investment (ROI)** • **Oversight and compliance** • Project readiness • Repeatability / scalability • **Stakeholder coordination** • **Technical merit**

FIGURE 3-1 Criteria that support a prioritization framework for infrastructure projects in the Gulf of Mexico region.
NOTE: Boldface indicates criteria identified as significant for infrastructure projects in the Gulf region.

RELATIVE WEIGHTING OF THE MACRO-CRITERIA

Participants completed handouts at each domain on Days 1 and 2, assigning 100 percentage points across the five macro-criteria, to indicate how they believe each should be weighted relative to one another when prioritizing projects.[2,3] Scoring across all scenarios and all domains suggests relatively close weighting. Overall, Resilience is weighted heavily relative to other macro-criteria. Project Governance and Economy are weighted slightly below Society and Environment. Figure 3-2 shows the aggregated results of all inputs. The box area for each macro-criterion reflects the middle two quartiles of scores from participants, and the whiskers reflect the first and last quartiles for each. Dots reflect outlier scores.

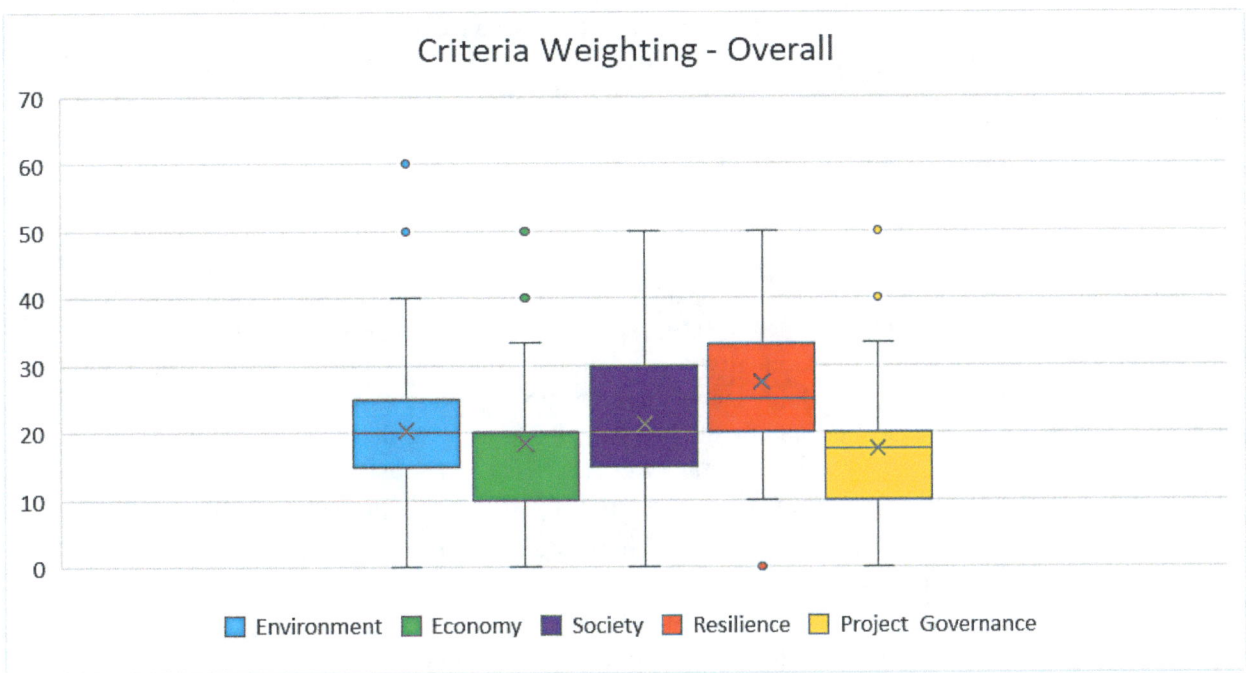

FIGURE 3-2 Relative weighting of macro-criteria by participants across both scenarios and all domains. X marks median values.

ELABORATION ON THE SUB-CRITERIA

Tables 3-1 through 3-5 offer descriptions of the sub-criteria in alphabetical order, not priority order. Readers may use these tables to gain further ideas or insights on how to interpret and apply the sub-criteria. Boldface sub-criteria were identified by participants as significant across all domains. These often also appear in other frameworks that serve to help prioritize investments to increase infrastructure resilience, even if the sub-criteria are understood or defined a bit differently across the various sources. Descriptions in the tables include perspectives from the workshop participants; the footnotes offer examples of how select sub-criteria appear in other well-known frameworks that informed development of the framework and design of the workshop.

---

[2] The handouts also asked whether participants had suggestions for additional macro-criteria categories. Only four handouts suggested additional macro-criteria, suggesting the initial five are sufficient.
[3] The workshop designers and the National Academies recognize that there are shortcomings of simple linear weighting schemes, but the process and results are more about prompting the discussions and identifying issues than the actual scores.

TABLE 3-1 Discussion of Each Sub-Criterion in the Environment Macro-Criterion, from Participants and from the Literature

Environment	
Advancement of scientific understanding of the environment	The extent to which projects support the advancement of the scientific understanding of the relevant environment should be considered when prioritizing projects. For example, participants commented that conducting baseline studies in areas where they are lacking is critical because measuring the impacts of incidents or of projects will be very difficult if there are no data or metrics against which to benchmark changes. They added that monitoring, mapping, assessment, analysis, and investments in new technologies are other ways to improve our understanding of the environment.
Biodiversity[a]	Participants commented that projects have not always considered their impacts on local species, but moving forward they must. For example, certain methods of de-oiling birds have proven to be ineffective and, in fact, caused more harm than good. Projects should consider and acknowledge these impacts and explain how they will be addressed. Projects should at minimum not adversely impact biodiversity, and ideally, they would enhance biodiversity. Some participants felt biodiversity should be considered a sub-component of climate change.
Climate Change[b]	Participants argued that to be prioritized, projects should address climate change, and they identified several ways for projects to satisfy this sub-criterion: • Projects should work to adapt infrastructure to projected changes in climate • Projects should improve knowledge of infrastructure sensitivities to climate change • Projects should account for both current and future impacts of climate change; they should also improve existing conditions • When modeling climate impacts, projects should assume worst-case projections and build infrastructure that will withstand these impacts, as well as account for uncertainty • Participants commented that when choosing between two climate change projects, the one that offers multiple benefits should take priority.
Degradation prevention	Participants suggested that projects should be prioritized if they focus on preventing environmental degradation in the first place, over projects that focus on restoration and recovery.
Do no harm/reduce harm to environment[c]	Participants suggested that projects should focus on reducing harm to the environment on a net basis first and if possible, account for unintended consequences second. They also observed, however, that sometimes, some harm to the environment must be done to support long-term sustainability and thrivability (e.g., employment), so that nuance and complexity should be kept in mind when evaluating projects.
Land use[d]	Participants noted that land use must be considered, possibly much more strongly than it has been in the past. Projects should keep people and infrastructure out of high-risk areas, either moving them or not putting them there in the first place; participants continued that any project that intends to move people, however, must require community leadership and buy-in so that no one is forced to leave their home. Additionally, projects should seek to preserve or even create communal and recreational land, and projects should address cascading impacts to land.

Maintenance	Participants explained that, similar to the "Persistence of efforts" sub-criterion below, the environment is not a "set-it-and-forget-it" issue. Given that some projects will take years to complete, they should account for long-term resource needs and maintenance costs.
Mitigation	Similar to the "Degradation prevention" sub-criterion above, participants commented that projects that mitigate impacts to the environment should be prioritized over those that simply respond to impacts.
Persistence of efforts	Participants explained that the environment typically receives the most attention after a disaster strikes, when there is visible impact or damage. To make real progress and truly improve environmental conditions, however, priority should go to projects that focus on long-term restoration after incidents and continuous preservation before incidents.
Riverine and coastal erosion[e]	Participants reported that riverine and coastal erosion are very important issues but are not studied enough, so projects should be prioritized that further the general knowledge of these topics. Projects should be prioritized that emphasize using natural barriers and green infrastructure to combat riverine and coastal erosion; if a project proposes gray infrastructure, it should consider how it will interact with natural barriers already in the area. Projects that support retrofitting and minimizing impacts from erosion should also be prioritized.
Safe living and working spaces	Participants suggested that projects should be prioritized if they emphasize the importance of equitably creating safe (hazard-free) living and working spaces, which are the foundation of a functioning community.
Sustainability	Participants suggested that projects should be prioritized if they take deliberate steps toward supporting environmental sustainability.
Water/air quality[f]	Participants reported that projects should be reviewed to ensure they are taking into account impacts to water and air quality at three levels: (1) now, (2) as a result of disasters, and (3) as a result of future impacts from climate change. They reported that current systems and infrastructure, especially related to the oil, gas, and petrochemical industries, are failing to take into account future impacts to water and air quality from climate change, and this must change. And similar to "biodiversity" above, some participants felt water/air quality should be considered a sub-component of climate change.

NOTES:

[a] The Standard for Sustainable and Resilient Infrastructure (SuRe®) includes biodiversity in its criteria for receiving certification that an infrastructure project follows Good International Industry Practice (GIIP) for sustainability and resilience in infrastructure. An example of one SuRe criterion for biodiversity, Biodiversity and Ecosystem Management, states that "the Project shall take actions to avoid negative impacts and maximise positive impacts on the conservation of Biodiversity such as species, habitats, (natural, modified, and/or critical), legally protected and internationally recognized areas, ecological corridors and ecosystems, which might arise from the Project." Another SuRe criterion for biodiversity, Biodiversity and Ecosystem Conservation, states that "the Project shall collaborate with state and local agencies in the protection and conservation of Natural Capital, Critical Habitats, ecosystems and species as recognised by the International Union for Conservation of Nature (IUCN) Red list." SuRe provides additional language for each of these criterion in its publication. See https://sure-standard.org/wp-content/uploads/2019/10/ST01_Normative_Standard_v1.1_clean.pdf.

[b] SuRe® includes climate change in its criteria for receiving certification that an infrastructure project follows GIIP for sustainability and resilience in infrastructure. One SuRe criterion for climate change, Climate Change Mitigation, states that "the Project shall avoid, or if not feasible, reduce project related Greenhouse Gas (GHG) emissions by assessing and implementing alternative solutions considered to be technically feasible and financially feasible throughout the lifecycle of The Project." Another SuRe criterion for climate change, Climate Change Adaptation, states that "the Project shall demonstrate its ability to withstand identified climate change risks and hazards in plausible scenarios throughout The Project's lifecycle." SuRe provides additional language for each of these

criterion in its publication. See https://sure-standard.org/wp-content/uploads/2019/10/ST01_Normative_Standard_v1.1_clean.pdf. The Federal Emergency Management Agency's (FEMA) Flood Mitigation Assistance Grant program also includes considerations of climate change in its scoring criteria: "… describe how the project will enhance climate adaptation and resilience, detail how the project is being responsive to the effects of climate change (such as sea level rise, increased rainfall, increased likelihood of flash flood due to wildfire, etc.) and/or other future conditions (population/demographic/land use, etc.), and cite data sources, assumptions, and models." See https://www.fema.gov/sites/default/files/documents/fema_nofo-fiscal-year-2021-flood-mitigation-assistance-grants.pdf.

[c] The no-harm principle obligates a state to prevent, reduce, and control the risk of environmental harm, as it relates to activities within their jurisdiction, to other states. See https://leap.unep.org/knowledge/glossary/no-harm-rule. This principle serves as the cornerstone of environmental law. For further information, see Benoit Mayer, "The relevance of the no-harm principle to climate change law and politics," *Asia Pacific Journal of Environmental Law* 19.1 (2016): 79–104.

[d] FEMA's Flood Mitigation Assistance Grant program includes the idea of land use, insofar as the degree to which repetitive loss (RL) and severe repetitive loss (SRL) properties are included in the "benefitting areas of the project" as a scoring criteria for its projects. See https://www.fema.gov/sites/default/files/documents/fema_nofo-fiscal-year-2021-flood-mitigation-assistance-grants.pdf. SuRe® includes land use in its criteria for receiving certification that an infrastructure project follows GIIP for sustainability and resilience in infrastructure: "the Project shall minimise the use of green-field land. The Project shall also employ technologies that reduce footprint and minimise impervious space. If the site may contain chemical, biological or radioactive contamination, the Project shall ensure that the site is adequately decontaminated prior to construction." See https://sure-standard.org/wp-content/uploads/2019/10/ST01_Normative_Standard_v1.1_clean.pdf.

[e] SuRe® includes soil restoration in its criteria for receiving certification that an infrastructure project follows GIIP for sustainability and resilience in infrastructure, specifically, "the Project shall promote sustainable use of soil through protection and restoration measures." SuRe includes this sub-criterion under its Land Use and Landscape category, and it provides additional language about this criterion in its publication. See https://sure-standard.org/wp content/uploads/2019/10/ST01_Normative_Standard_v1.1_clean.pdf.

[f] SuRe® includes air and soil pollution in its criteria for receiving certification that an infrastructure project follows GIIP for sustainability and resilience in infrastructure: "the Project shall be designed, built, operated and decommissioned in a way that avoids or minimises the pollution of air and soil and avoids the transfer of Pollution from air to soil or from soil to air." It also includes water pollution: "the Project shall avoid or when avoidance is not possible, minimize the Pollution of water and the transfer of Pollution from water to other resources (land, air, etc.)." SuRe provides additional language about this criterion in its publication. See https://sure-standard.org/wp-content/uploads/2019/10/ST01_Normative_Standard_v1.1_clean.pdf.

TABLE 3-2 Discussion of Each Sub-Criterion in the Economy Macro-Criterion, from Participants and from the Literature

	Economy
Business continuity	Participants commented that projects should be prioritized if they include the development of plans to provide small businesses with the means to continue operations after incidents. Participants explained that having a program in place before an incident will help keep the economy functioning, reducing job losses, business closures, and the loss of the tax base.
Commerce support	Participants suggested that a project should be prioritized if it specifically focuses on restarting, driving, or supporting commerce. Some participants felt uninterrupted or enhanced commerce is one of the, or the, most important aspect of the economy criterion. Also, projects should be prioritized if they support the microeconomies and small businesses in the region, and minimally, all projects should address their impacts on the local or regional microeconomies and businesses.

Continuity/ recovery of economic activity[a]	Participants suggested that projects should be prioritized if they clearly support rapid resumption of economic activity or continuation of markets. Some participants suggested that it is not effective to just focus on employment or the tax base; it should be economic activity from a holistic perspective.
Cost-effectiveness[b]	Participants suggested that project proposals should specifically address how the projects are cost-effective, and cost-benefit assessments should account for noneconomic impacts in addition to economic impacts.
Economic development[c]	Participants suggested that projects should encourage economic development to whatever extent possible, even if those benefits will not be realized until years from now. They added that projects should strive for equity in economic development and prioritize the needs of underserved communities.
Employment[d]	Participants commented that projects must, first, prevent job loss and, second, create jobs. And even better, they suggested that a project should strive to create diverse employment opportunities in the region to maximize opportunities for populations with various backgrounds and skills.
Equity	Participants suggested that equity of outcomes should be an overarching theme addressed by all projects, but it should be specifically assessed under the Economy and Society macro-criteria. Here, "equity" should focus on ensuring equity of financial outcomes as a prioritization criterion.
Financial need	Participants commented that state governments, local governments, and organizations in the regions have already maximized what they can do with their existing funds. Projects should be prioritized if they are addressing unmet needs of these agencies.
Individual support	Some participants thought projects should be prioritized if they address the immediate needs of individuals. They explained that the local community has needs, and the community must see short-term benefits from the projects they often hear about. Others saw this as a part of the Employment and/or Economic development sub-criteria.
Labor rights and working conditions[e]	Participants reported that projects that involve labor laws or otherwise directly improve working conditions should be prioritized. This might include, for example, projects designed to prevent worker burnout or worker fatigue, which can lead to adverse health impacts across the community.
Multiple impacts	Participants commented that projects that address multiple or complex impacts should be prioritized over those that address a more narrow impact.
Proximity to major economic and/or residential centers	Participants suggested that to maximize positive impacts for as many communities as possible, projects should be located near major economic and/or residential centers, if feasible and safe to do so.
Spending drivers	Similar to "Continuity/recovery of the economy," above, participants suggested that projects should be prioritized if they take deliberate steps to drive spending—by the government, by businesses, or by residents—because spending jumpstarts the economy and should be encouraged.
Tax base	Participants explained that keeping the tax base robust is key to supporting services funded by state and local government agencies that are very important to the community, so projects that support the tax base should be prioritized. Further, this is important because there are downstream effects—participants commented that maintaining a healthy tax base and thus resultant services helps encourage people to not leave the area, keeps unemployment down, and prevents other adverse impacts to the economy, and in turn to the people.

NOTES:

a Several frameworks in use today promote the conduct of cost-benefit analyses to inform project selection. There is more information related to cost-benefit analysis and cost-effectiveness in general under the next sub-criterion, but one element of cost-benefit analysis relates directly to continuity/recovery of economy activity. "Benefits" of a potential project often include measures that support loss reduction, which supports recovery of the economy. The National Institute of Standards and Technology (NIST) Community Resilience Economic Decision Guide for Buildings and Infrastructure Systems describes an approach to cost-benefit analyses that addresses this directly. See https://nvlpubs.nist.gov/nistpubs/SpecialPublications/NIST.SP.1197.pdf.

b Cost-effectiveness is a predominant component of several other existing frameworks. See FEMA's Hazard Mitigation Assistance grant program (https://www.fema.gov/grants/guidance-tools/benefit-cost-analysis); the HUD [U.S. Department of Housing and Urban Development] Rebuild by Design (CPD-16-06) guidelines (https://www.hud.gov/sites/documents/16-06CPDN.PDF); and the USACE [U.S. Army Corp of Engineers] "Cost Effectiveness for Environmental Planning: Nine EASY Steps," for detailed processes to conduct cost-effectiveness and cost-benefit analyses. See https://www.iwr.usace.army.mil/Portals/70/docs/iwrreports/94-PS-2.pdf

c The Standard for Sustainable and Resilient Infrastructure (SuRe®) includes indirect/direct economic development enabled by the project in its criteria for receiving certification that an infrastructure project follows Good International Industry Practice (GIIP) for sustainability and resilience in infrastructure: "the Project shall contribute to local socioeconomic development priorities throughout its life cycle and beyond, aligned with local and national development goals." SuRe provides additional language about this criterion in its publication. See https://sure-standard.org/wp-content/uploads/2019/10/ST01_Normative_Standard_v1.1_clean.pdf.

d SuRe® includes direct employment and training in its criteria for receiving certification that an infrastructure project follows GIIP for sustainability and resilience in infrastructure: "the Project shall hire people from the local communities as workers, professionals and in managerial positions during construction and operation of the Project." SuRe provides additional language about this criterion in its publication. See https://sure-standard.org/wp-content/uploads/2019/10/ST01_Normative_Standard_v1.1_clean.pdf.

e SuRe® includes labor rights and working conditions as a category in its criteria for receiving certification that an infrastructure project follows GIIP for sustainability and resilience in infrastructure. It breaks this category down into nine sub-criterion, including: (1) employment policy; (2) ensuring rights to association and collective bargaining; (3) nondiscrimination; (4) forced labor and child labor; (5) occupational health and safety; (6) employee grievance mechanism; (7) working hours and leave; (8) fair wages and access to employee documentation; and (9) retrenchment. See https://sure-standard.org/wp-content/uploads/2019/10/ST01_Normative_Standard_v1.1_clean.pdf.

TABLE 3-3 Discussion of Each Sub-Criterion in the Society Macro-Criterion, from Participants and from the Literature

Society	
Access to information/awareness	Participants explained that project proposers and managers should make the approach, methodology, and anticipated impacts of projects transparent. They should present this information in a user-friendly and consumable format so all stakeholders have awareness of what is taking place or expected to take place. They should extend access to this information to the general public whenever possible and appropriate. Participants added that this sharing of information should promote equal access to information and resources, and it should offer the community tools to educate themselves. This will ultimately build and maintain trust. Further, given the current trends on mis-, dis-, or malinformation (MDM), some participants believed a project should have a plan in place to counter the spread of MDM.

Benefits go to underserved communities[a]	Participants stated that priority should go to projects that benefit underserved communities and provide outcomes that are fair and equitable. They added that projects should address long-term systemic problems of these underserved communities.
Community engagement[b]	Participants suggested that this is possibly the most important sub-criterion of all; that a project will never succeed without deep and sustained community engagement. This includes partnership between project managers and the community and involving local leadership in project-related planning and execution. Other participants suggested that projects should also leverage volunteer support and crowd sourcing, where possible. Also, if the community is engaged and there is buy-in, it lessens the need for conflict resolution later.
Connections	Participants suggested projects should be prioritized that help restore connections among the impacted population, because connections support a return to normalcy and/or stability. These include, for example, projects that restore communications so that people can "connect" with their friends and family.
Cultural value	Participants commented that projects should be respectful of land, flora, and/or fauna that are culturally valuable to communities. Minimally, project proposals should explain how they will do no harm to these natural cultural assets.
Education	Participants explained that many people in the Gulf region feel that the government does things to people, and not for people, so projects should include public education. Participants suggested projects should be prioritized if they educate the public about how to increase their own resilience and/or if they educate the public about how to reduce their own risk, for example, forcibly removing people out of high-risk areas is very difficult, but if the impacted people understand the benefits, they may take voluntary action on their own. Finally, people in the region already volunteer their time to supplement response operations, but if projects include education about these activities, their volunteering and thus the response will be even more effective.
Equity and inclusion[c]	Participants suggested that equity and inclusion should be a priority, and it should be prioritized over cost. Projects should strive for equity in jobs and resource sharing and/or offset impacts of local industries if they are currently harmful to communities. Because equity can be difficult to measure, project proposals should describe specific and quantifiable impacts on different communities and how inclusion will be approached. Unlike other sub-criteria, participants believed equity and inclusion cannot be assessed simply as a binary choice given the diversity in communities that exists across the Gulf region.
Health and welfare[d]	Participants suggested that project proposals should include specific metrics to measure impacts of the project on human health and that projects should also monitor impact on human health throughout project implementation.
Public safety/order[e]	Participants explained that people must feel safe first before any other initiatives can be effective, so projects should be prioritized that improve safety. This includes preventing loss of life and/or reducing the risk of death. It also includes enhanced disaster response and, specifically, support for mental health. Several participants explicitly stated that this sub-criterion is not about prioritizing law enforcement, but about general community welfare and safety.

Public trust	Participants suggested projects should be prioritized if they help to increase the public's understanding of governmental actions and work to increase public confidence in government. This might be achieved, for example, through projects that explain how they will support increased transparency around government activities, how they will utilize community engagement to build public trust, how they will create partnerships with trusted voices in the community, and how they will promote core social values (e.g., use of the environmental, social, and governance (ESG) framework).
Return to stable society/normalcy	Participants suggested that projects must first and foremost support a return to a stable society or normalcy after an incident. Projects that help promote rapid and effective returns to normalcy should be prioritized until survivability is no longer at risk.
Social and political acceptability	Participants explained that a project should include advice and consent from the people it will affect. Further, buy-in from marginalized groups who are already wary of governmental intrusion is necessary for long-term project success; project proposals must account for achieving this buy-in.
Survivability (e.g., water, basic needs)	Similar to "Return to stable society/normalcy," above, participants reported that basic survival is still at risk immediately after an incident, and so projects that directly support survival should be prioritized. This includes access to clean drinking water, functioning wastewater systems, communications, electricity, and housing.

NOTES:

[a] The Federal Emergency Management Agency's (FEMA) Flood Mitigation Assistance Grant program includes the Centers for Disease Control and Prevention's (CDC) Social Vulnerability Index (SVI) as a scoring criteria. More specifically "projects that benefit area(s) with an overall SVI score of 0.7501 or greater per CDC's Social Vulnerability Index will [be] eligible for this point priority…. In the event multiple census tracts are included in an area benefiting from the project, FEMA will consider the highest SVI score." See https://www.fema.gov/sites/default/files/documents/fema_nofo-fiscal-year-2021-flood-mitigation-assistance-grants.pdf.

[b] The Standard for Sustainable and Resilient Infrastructure (SuRe®) includes stakeholder engagement as a category in its governance criteria for receiving certification that an infrastructure project follows Good International Industry Practice (GIIP) for sustainability and resilience in infrastructure. Based on participant feedback during the workshop, this concept may be relevant to both the society and the project governance macro-critieria. SuRe breaks this category down into three sub-criterion, including (1) stakeholder identification and engagement planning; (2) engagement and participation; and (3) public grievance and customer feedback management. See https://sure-standard.org/wp-content/uploads/2019/10/ST01_Normative_Standard_v1.1_clean.pdf.

[c] FEMA's Guide to Expanding Mitigation, Making the Connection to Equity, suggests the following ways to measure equity successes in mitigation initiatives, for example: (1) at-risk populations lend expertise and have agency in the hazard mitigation process; (2) barriers to participating in mitigation activities are removed, and training, language access, transportation, meals and/or childcare are provided; (3) investment takes place in traditionally underserved communities; (4) race is no longer a determining factor of risk; and (5) indicators of social vulnerability and environmental burden are assessed alongside hazards in mitigation plans and are used to target outreach or risk reduction projects. The publication includes additional considerations. See https://www.fema.gov/sites/default/files/documents/fema_mitigation-guide_equity.pdf. Arup International Development's City Resilience Framework identifies inclusion as a quality of a resilient system: "Inclusion emphasizes the need for broad consultation and engagement of communities, including the most vulnerable groups. Addressing the shocks or stresses faced by one sector, location, or community in isolation of others is an anathema to the notion of resilience. An inclusive approach contributes to a sense of shared ownership or a joint vision to build city resilience." See https://www.rockefellerfoundation.org/wp-content/uploads/City-Resilience-Framework-2015.pdf.

[d] SuRe® includes both management of public health and safety risks and delivery of public health and safety benefits in its criteria for receiving certification that an infrastructure project follows GIIP for sustainability and resilience in infrastructure. See https://sure-standard.org/wp-content/uploads/2019/10/ST01_Normative_Standard_v1.1_clean.pdf.

[e] SuRe combines public health and safety into the same two sub-criteria.

TABLE 3-4 Discussion of Each Sub-Criterion in the Resilience Macro-Criterion, from Participants and from the Literature

Resilience	
Adaptability[a]	Participants acknowledged that adaptability is a more complex sub-criterion than many of the others, but that to lead to resilience, projects must be designed to adapt to future conditions, even if those conditions are not yet known. Changing weather patterns, sea level rise, and other impacts from climate change may adversely affect the outcome of projects based on current conditions, and so projects must be designed so that they can adapt to worsening conditions. The inclusion of green infrastructure in project design would also be appropriate here.
Exposure reduction	Participants suggested that projects that support exposure reduction of either people or infrastructure should be prioritized. This might include encouraging people to move out of flood zones or away from chemical plants. Prioritize projects that would decrease exposure to rising sea level for both people and infrastructure. Participants added that projects should use (or change) regulations and enforcement to address abandoned and exposed infrastructure before they become hazards.
Focus on preparedness and response	Participants explained that not only does the Gulf region have experience with natural hazards but it expects to have more frequent and forceful weather impacts to its infrastructure in the future. Participants further explained that oil spills have happened in the region before, and they will continue to happen. Therefore, projects should be prioritized if they do more than just repair damage—they should help prepare infrastructure and society for these events and also improve the state and local responses to those events.
Hardening of infrastructure[b]	Participants suggested that projects should be prioritized if they include solutions to mitigate the impacts from a disaster, solutions to adapt flexibly to disasters, measures of redundancy, and multiple lines of defense against threats and hazards. Projects should take more natural approaches to infrastructure investment (i.e., green infrastructure), if feasible; if not feasible, gray infrastructure projects that allow green infrastructure to be included should be prioritized.
Information management and communications	Participants explained that projects should directly address clearly how they will manage information and support transparent communications. Projects should be prioritized that include a thoughtful communications plan that offers an explanation of trade-offs and benefits of the project to the local community.
Interdependence with other systems[c]	Participants explained that interdependence among infrastructure systems and networks is positive and something projects should strive to support. That being said, to prevent larger system failure, participants noted that projects must address interdependencies and explain how they will mitigate potential adverse impacts from a disaster and ensure operability if a system is inaccessible. Any new project should minimally ensure that its implementation does not harm other projects in the system or area and, ideally, leverage and compliment them.
Long-term impact	Participants suggested that projects that can illustrate long-term benefits should be prioritized over those that cannot or do not.

Planning and preparedness (prevention focus)	Participants suggested that projects should be prioritized if they include planning to prevent known hazards from happening in the first place or prevent major impacts if they do occur. Better yet, projects that include expansive and creative consideration of future risk versus historical risk should receive priority consideration.
Rapid deployment for response	Participants suggested that projects enhancing the ability to respond rapidly to a disaster should be prioritized, because they believed part of resilience is the ability to quickly deploy and get to work responding without delay.
Redundancy[d]	Participants commented that project proposals should include a clear explanation of fail-safe alternatives they are including to help with quicker recovery of systems after a disaster and minimize negative or cascading impacts from that disaster. From a sustainability perspective, priority should go to projects that support systems with more than one purpose. Projects should also address workforce redundancy (possibly achieved through training) to ensure continuity of operations.
Supportive of the system/network[e]	Participants suggested that projects should deliberately address how they are supplementing or complementing systems or projects already in development and not work against them or unnecessarily replicate them. Some participants considered this sub-criterion as complimentary of, but more comprehensive than, "Interdependence with other systems," listed above.
Understanding AND action	Participants suggested that projects that both increase knowledge and incite action should be prioritized over those that just do one or the other. Increasing knowledge might include enhancing situational awareness or understanding of disaster impacts through improved data collection, analysis, or research. This increased understanding should automatically drive related action.

NOTES:

[a] Arup International Development's City Resilience Framework (Arup) identifies flexibility as a quality of a resilient system and defines it similarly to the spirit of the adaptability sub-criterion: "Flexibility implies that systems can change, evolve and adapt in response to changing circumstances.... Flexibility can be achieved through the introduction of new knowledge and technologies, as needed. It also means considering and incorporating indigenous or traditional knowledge and practices in new ways." See https://www.rockefellerfoundation.org/wp-content/uploads/City-Resilience-Framework-2015.pdf.

[b] Arup identifies robustness as a quality of a resilient system and defines it similarly to the hardening of infrastructure sub-criterion: "Robust systems include well-conceived, constructed and managed physical assets, so that they can withstand the impacts of hazard events without significant damage or loss of function. Robust design anticipates potential failures in systems, making provision to ensure failure is predictable, safe, and not disproportionate to the cause." It provides additional language in its publication. See https://www.rockefellerfoundation.org/wp-content/uploads/City-Resilience-Framework-2015.pdf. The Federal Emergency Management Agency's (FEMA) Flood Mitigation Assistance and Building Resilient Infrastructure and Communities grant program includes "incorporation of nature-based solutions" as a scoring criteria. See https://www.fema.gov/sites/default/files/documents/fema_nofo-fiscal-year-2021-flood-mitigation-assistance-grants.pdf; https://www.fema.gov/sites/default/files/documents/fema_nofo-fiscal-year-2021-building-resilient-infrastructure.pdf and https://www.fema.gov/sites/default/files/documents/fema_nofo-fiscal-year-2021-building-resilient-infrastructure.pdf.

[c] Arup identifies integration as a quality of a resilient system: "integration and alignment between city systems promotes consistency in decision-making and ensures that all investments are mutually supportive to a common outcome. Integration is evident within and between resilient systems, and across different scales of their operation. Exchange of information between systems enables them to function collectively and respond rapidly through shorter feedback loops throughout the city." See https://www.rockefellerfoundation.org/wp-content/uploads/City-Resilience-Framework-2015.pdf.

[d] Arup identifies redundancy as a quality of a resilient system and defines it as "spare capacity purposely created within systems so that they can accommodate disruption, extreme pressures or surges in demand. It includes

diversity: the presence of multiple ways to achieve a given need or fulfil a particular function. Examples include distributed infrastructure networks and resource reserves. Redundancies should be intentional, cost-effective and prioritised at a city-wide scale, and should not be an externality of inefficient design." See https://www.rockefellerfoundation.org/wp-content/uploads/City-Resilience-Framework-2015.pdf.

[e] Arup's criteria address interdependence with other systems.

TABLE 3-5 Discussion of Each Sub-Criterion in the Project Governance Macro-Criterion from Participants and from the Literature

Project Governance	
Accounting for cyber and physical security implications	Participants suggested that projects must deliberately consider and acknowledge any cyber or physical security implications, and if such implications do exist, projects should include actions to address them.
Alignment with existing plan	Similar to the "Supportive of the system/network" sub-criteria under Resilience, participants suggested that since many plans are already in place and many projects are already underway, new projects should be prioritized if they leverage work already done or utilize teams already formed. It would be acceptable for these plans or teams to require updates to account for equity and fairness, however.
Community inclusion[a]	Many participants believe projects must collect inputs from the community on their design and execution, in order to be successful and/or effective. Projects should include mechanisms for sharing information with the local population and educating the community about potential benefits they can expect from a project. Projects should aim to lower bureaucratic barriers as much as possible for the community to get involved effectively and successfully. Some workshop participants believe this sub-criterion is a subset of "stakeholder engagement."
Data and information sharing[b]	Some participants argued that, first, a project must not impede current information-sharing efforts. Second, a project should illustrate how it will create and deliver its information in a user-friendly, easy-to-consume way. Data and information sharing should specifically include the sharing of best practices and lessons learned from past experiences.
Feasibility[c]	Some participants believed a project proposal must specifically address why it is feasible. Minimally, the proposal must make clear that it will follow existing codes and how it will do that. Some participants did not think feasibility should be a part of project governance because it should be worked out ahead of time, but they did not specify when or where.
Outcomes focus—effectiveness, return on investment (ROI)[d]	Participants suggested that projects should include mechanisms to assess if they are generating the impacts and benefits promised or expected. Others suggested metrics would be useful here and so projects should offer mechanisms to clearly measure both their progress and their effectiveness. Others felt current frameworks for prioritizing projects should be leveraged, as they offer reasonable ways to evaluate outcomes appropriately.
Oversight and compliance[e]	Participants reported that projects must include some explanation of how they will comply with relevant laws, codes, and other norms.
Project readiness	Participants suggested that projects should be prioritized if proposals clearly address project readiness, which might include, for example, availability of trained staff or identification of proven experience securing relevant permits.
Repeatability/scalability	Participants suggested that projects should be prioritized if proposals make explicit how the projects can be both repeatable and scalable.

Stakeholder coordination[f]	Participants suggested that projects should be prioritized if they clearly identify project stakeholders and have a plan in place to engage them regularly for information-sharing and collaboration purposes.
Technical merit[g]	Participants suggested that technical merit should be assessed in some way, and that project proposals should address how they will manage project risk.

NOTES:

[a] Arup emphasizes the need for broad consultation and engagement of communities, including the most vulnerable groups, to create resilience. The Standard for Sustainable and Resilient Infrastructure ((SuRe®) includes engagement and participation as sub-criteria. See https://sure-standard.org/wp-content/uploads/2019/10/ST01_Normative_Standard_v1.1_clean.pdf. The Federal Emergency Management Agency's (FEMA) Building Resilient Infrastructure and Communities grant program includes "outreach activities" as part of its evaluation criteria. See https://www.fema.gov/sites/default/files/documents/fema_nofo-fiscal-year-2021-building-resilient-infrastructure.pdf.

[b] SuRe® includes public disclosure in its criteria for receiving certification that an infrastructure project follows Good International Industry Practice (GIIP) for sustainability and resilience in infrastructure. It explains that minimally a project must disclose all information required to remain in compliance with applicable laws, and then it offers a list of additional types of project information that should be disclosed even if not required by law. See. https://sure-standard.org/wp-content/uploads/2019/10/ST01_Normative_Standard_v1.1_clean.pdf.

[c] SuRe® includes project team competence in its criteria for receiving certification that an infrastructure project follows GIIP for sustainability and resilience in infrastructure, described this way: "for the construction and the operation phases of the Project, the Project Team(s), including those of its Direct Contractors, shall consist of skilled and experienced professionals qualified to fulfil their tasks and responsibilities and are appointed based on merit via a transparent recruitment process." See https://sure-standard.org/wp-content/uploads/2019/10/ST01_Normative_Standard_v1.1_clean.pdf.

[d] SuRe® includes results orientation in its criteria for receiving certification that an infrastructure project follows GIIP for sustainability and resilience in infrastructure, described this way: "the Project shall define goals and objectives with regard to the primary purpose of The Project and define Key Performance Indicators (KPIs) accordingly." SuRe provides additional language about this criterion in its publication. See https://sure-standard.org/wp-content/uploads/2019/10/ST01_Normative_Standard_v1.1_clean.pdf.

[e] SuRe® includes legal compliance and oversight in its criteria for receiving certification that an infrastructure project follows GIIP for sustainability and resilience in infrastructure, described this way: "the Project Owner shall ensure that the Project complies with the applicable laws and regulations throughout its life cycle. Applicable laws and regulations shall include local (municipal and regional), national legal, regulatory and administrative requirements as well as applicable international law and indigenous rights." See https://sure-standard.org/wp-content/uploads/2019/10/ST01_Normative_Standard_v1.1_clean.pdf. FEMA's Flood Mitigation Assistance Grant program does not necessarily include this as a sub-criterion, but it does conduct monitoring and oversight as part of its management process; it makes "site visits or conducting desk reviews to review project accomplishments and management control systems to review award progress and to provide any required technical assistance." See https://www.fema.gov/sites/default/files/documents/fema_nofo-fiscal-year-2021-flood-mitigation-assistance-grants.pdf.

[f] FEMA's Building Resilient Infrastructure and Communities grant program includes "leveraging partners" as an evaluation criteria, explained this way: "the project subapplication incorporates partnerships (e.g., state, tribal, private, local community, etc.) that will ensure the project meets community needs, including those of disadvantaged populations, and show the outcome of those partnerships (e.g., leveraging resources such as financial, material, and educational resources, coordinating multi-jurisdictional projects, heightened focus on equity related issues, etc.)" See https://www.fema.gov/sites/default/files/documents/fema_nofo-fiscal-year-2021-building-resilient-infrastructure.pdf. Also, as described in footnote 27 for "Community engagement" under the Society macro-criterion, above, SuRe includes three criteria related to stakeholder engagement: (1) stakeholder identification and engagement planning, (2) engagement and participation, and (3) public grievance and customer feedback management. While the first was particularly relevant to the "Community engagement" sub-criterion, the second is particularly relevant to this, the "Stakeholder engagement" sub-criterion under Project Governance. See https://sure-standard.org/wp-content/uploads/2019/10/ST01_Normative_Standard_v1.1_clean.pdf.

g SuRe® includes risk management in its criteria for receiving certification that an infrastructure project follows GIIP for sustainability and resilience in infrastructure, described this way: "the Project shall make regular and comprehensive assessment and management of current and future risks; including natural hazards, environmental, social, governance, policy, technological and economic risks relating to the construction and operation phases of The Project. Risks assessed shall include those caused by third parties' actions that have an impact on The Project's area of influence." See https://sure-standard.org/wp-content/uploads/2019/10/ST01_Normative_Standard_v1.1_clean.pdf.

Part 4

Next Steps

The workshop concluded with a hotwash session and discussion of next steps. Each participant responded to any combination of the following questions that they preferred:

1. Did you learn anything new this week? What did you learn?
2. What did you find particularly valuable about this week?
3. What should we do differently if we hold this type of workshop again?
4. Is there any new action you will take based on this workshop?
5. What next steps or additional activities would be most useful to you, your organization, and this overall process?

This part provides some general observations about the process and an overview of the next steps suggested by workshop participants through their responses to the questions above. Appendix C provides an overview of next steps to develop the prioritization framework based on best practices from the literature.

It was evident throughout the workshop that

- the approach was inclusive and can be broadened further to achieve consensus, when needed;
- there was a collective commitment to the workshop objectives and desire work toward effective implementation of infrastructure investment decisions;
- the knowledge and expertise of the participants in multiple domains was a critical element for success; and
- the process enables iteration and refinement, which provides a template for actions.

Circulate the Proceedings

Multiple participants expressed interest in reviewing the materials that are compiled as a result of this workshop, for their own reference, to share with colleagues, and to understand the action(s) the National Academies of Sciences, Engineering, and Medicine will take next.

Communicate with the White House

Multiple participants urged that the National Academies leverage the work done during this workshop and follow up with the White House. Follow-up should take place in time to influence the execution of the Infrastructure Investment and Jobs Act (IIJA, P.L. 117-58). It should include the key themes and takeaways that emerged from this workshop or take the form of a detailed checklist of criteria to be used for project evaluation. It should emphasize that project development and prioritization must be interdisciplinary.

Conduct Follow-on Workshops

Some participants suggested that the National Academies hold another event to transition from the higher-level issues identified during this workshop to more concrete, discrete ideas and projects that might support infrastructure resilience. Others suggested that the National Academies hold this same workshop again with different groups of experts, particularly with more state and local officials. Additionally:

- Multiple participants commended the methodology that was implemented through this workshop, describing it as rigorous and well planned, and suggested the same methodology be replicated moving forward.
- Many participants believed the diverse group of well-informed and committed attendees helped make this workshop so successful—they learned from one another and gained exposure to work going on outside of their own fields. They commented that similar types of invitation lists should be developed for future workshops.
- Several participants offered to help get their colleagues involved or to provide follow-up information for future workshops.

Consider if the National Academies Should Take a More Active Role

One participant suggested that the National Academies may want to become an "active buyer" in infrastructure resilience, developing a research plan for review by workshop participants and shaping and guiding future projects, rather than playing a more passive role.

Move Beyond a Checklist

Another participant suggested that the National Academies turn the work done in this workshop—specifically, the prioritization criteria—into a useful algorithm, moving beyond development of just a series of checkboxes into something even more useful.

A

Takeaways and Observations by Domain

This appendix provides key participant comments, takeaways, and observations by domain, by scenario, and by exercise from Days 1 and 2. These takeaways (see Tables A-1 through A-5) map to the scenario-specific takeaways and observations outlined in the body of this proceedings and ultimately drive the workshop-wide takeaways highlighted in the Summary.

All of the takeaways reflect comments and discussions at the workshop. Takeaways are presented as having been said by workshop participants. Readers should not regard these as carrying the weight of recommendations from the National Academies of Sciences, Engineering, and Medicine. The views cited are not necessarily consensus views of the group, and the group was not composed to meet National Academies standards for study committees that make consensus findings and recommendations. The takeaways do reflect key ideas presented or discussed by one or more workshop participants, so the material presented here could be considered suggestions coming from informed individuals in the process.

TABLE A-1 Overview of Workshop-wide Key Takeaways

Necessity of community engagement
• Emphasizing local leadership
• Promoting community involvement
• Improving communications and transparency
Call to overcome long-standing obstacles
• Moving from survivability to thrivability
• Focusing on prevention in addition to response
• Incorporating resilience and equity consistently
Need to move into the future
• Building up data and analysis
• Embracing technology and modernization
• Balancing regulations versus incentives carefully

TABLE A-2 Key Takeaways by Domain, by Scenario, and by Exercise: Petrochemical Industry Functions

Petrochemical Industry Functions	
Hurricane Scenario	
Project Brainstorming (P&P Exercise)	Participants commented: • Formulating projects that will increase infrastructure resilience requires **evaluating risk tolerance and trade-offs**: what does the community seek to avoid (e.g., storm surge above a particular height at a specific type of infrastructure) and how are stakeholders willing to pay to buy down that risk? • Investing in **retail fuel station resilience** is important. This includes prioritizing and requiring backup power generation and making stations "anchor institutions," for priority power restoration, but it is complicated by the complex governance network of fuel stations and varying regulations across states. • There would be benefit from petroleum industry owners **upgrading production and refining infrastructure** (e.g., use "smart tanks" capable of self-monitoring, auto shutoff, remote inspection after event), which would mitigate fallout and limit downtime during a with-notice event, but there is no business reason for them to do so. This has **stick (regulatory) and carrot (incentive) implications**; prudent use of both could be good for both industry and the public sector. • Participants **cautioned against overreliance on the private sector** to address hurricane-driven adverse impacts without incentivizing actions from the government. • There will likely always be a shortage of critical supplies during a response; therefore, it is useful to **decide beforehand who gets priority**. This includes **preplanning local and national priorities**, both to limit where essential resources are allocated and so that national requirements for recovery, which local leaders may not even be aware of, do not derail response operations once they are underway. For example, the East Coast depends on fuel from the Gulf region, and that has to be accounted for when planning response and recovery operations.
Project Prioritization (Prioritization Exercise)	Participants commented: • **Two types of resilient infrastructure investments are useful**: (1) ones that make things better in the **short term** and (2) ones that address the root problem(s) in the **long term**. The shorter-term fixes help people deal with the effects of events, even if marginally. Also, people need to see short-term successes because the long-term solution may take a very long time to implement. • It is beneficial to focus on preventing contamination in the first place, rather than just improving cleanup. Participants emphasized **enhancing storage tank resilience and secondary tank protections**. • Getting back to normal—in oil production, refining, transport, retail fuel operations—is important because it benefits so much: individuals, society, and the economy. This thinking favors projects that support an **expedited response** to the event, but also prioritization of key activities outside the impact area (e.g.,

APPENDIX A

	ensuring the availability of fuel in other parts of the country) for the sake of the economy.
	• Projects that **increase understanding of the damage petrochemical incidents do** to the environment rather than just projects that contain the damage would be beneficial. If the likely consequences or impacts to the environment are better understood, efforts can be focused more effectively to build back smarter.
	• **Using new technologies to their best advantage** is helpful, such as taking advantage of a**dvances in sensor technology** to allow the monitoring of flooded roadways and collecting samples in the middle of the ocean during a hurricane, or by creating smart fuel storage tanks.
	• Driving action from the petroleum industry after an event requiring cleanup **via sticks and carrots is tricky** and requires very smart governance. Environmental, social, and governance regulations or best practices seem promising, but there is debate as to whether the government is the right entity to develop the **regulations or document the best practices**, or if it is better to get support from other agencies.
Protracted Oil Spill Scenario	
Project Brainstorming (P&P Exercise)	Participants commented: • **Smarter regulation** of the petroleum industry will likely support a more resilient Gulf. This includes stricter permitting and stronger government enforcement authorities and expertise to compel the responsible party to clean up in a timely manner. • **Oversight of legacy and abandoned infrastructure** is another item that could be addressed through stricter permitting. The cost of maintaining and servicing aging petroleum infrastructure is a real challenge and a reason some owners walk away, resulting in ongoing public liabilities. • **R&D investments in petroleum infrastructure–related technology**, which would prevent and better contain spills, are valuable investments. • To improve response, there is a need to **determine, with greater accuracy and more authority, information about a spill**, such as the volume and impact of the spill, in ways that cannot be challenged by competing interests. • **Increase input from local leadership into response operations**. Bringing in local leaders will not only increase trust in the response effort but can also help encourage that local priorities and concerns are taken into account. • **The Oil Spill Liability Trust Fund (the "cleanup fund") would benefit from further refinement** to optimize the use of funds. For example, funds could be shared across federal agencies and used more flexibly to address local concerns and priorities, including R&D projects. • Given that infrastructure owners are typically responsible for cleanup after an incident, **owners have lessons learned that can be shared**. Currently, there is no way of organizing or compelling this sharing and plenty of reasons for owners not to share (e.g., concerns of disclosing information that might provide a competitive advantage), which means future cleanup efforts do not necessarily benefit from past response experiences—both successes and failures.

Project Prioritization (Prioritization Exercise)	Participants commented: • Useful areas for attention include **preventing spills in the first place** and mitigating spills if and when they occur by **requiring more of petroleum industry owners** through stricter permitting, including requirements for recovery plans, better monitoring, and funds set aside for cleanup. This benefits the economy, society, and environment by decreasing the likelihood of a spill, increasing the likelihood that the responsible party has plans and capabilities needed to contain and clean up the spill, increasing funding for addressing environmental degradation, and increasing public trust in the government's ability to manage or oversee oil spill responses. • **Being resilient means responsibly spending money over time**, not only in reaction to a big crisis. • It would be helpful to prioritize **more studies of potential risks**, such as landslides, and invest in predictive modeling so that permitters and industry can weigh the benefits of new drilling projects against the potential costs of a spill to society. It would be useful if **predictive modeling** considered multiple hazards occurring at the same time, to better anticipate future impacts. • There are still benefits to be gained, however, **if the speed of cleanup and recovery improves**, for multiple reasons: restoring the environment, the recovery of the local economy, and increasing trust in government response. • Public awareness and understanding of oil spill response is a special challenge. Government and industry stakeholders can **build and maintain trust with the public through clear communications** that do not add to the confusion. The government would benefit from faster, more detailed communications with the public to prevent misinformation. Better, faster public information sharing would be useful year-round. • **Remove aging and abandoned infrastructure** would help reduce risk and prevent future environmental degradation and pollution.

TABLE A-3 Key Takeaways by Domain, by Scenario, and by Exercise: Other Infrastructure Functions

Other Infrastructure Functions	
Hurricane Scenario	
Project Brainstorming (P&P Exercise)	Participants commented: • After a storm, **power should be restored as quickly as possible, followed by communications.** Local communities often have priority power restoration lists established, and they would benefit from preparing lists for communications restoration too. It would be useful to **map out key infrastructure system data**, consolidated into a central repository, and accessible by those who need it. • Restoring cell service is important. It enables communications and restores community confidence. A way to facilitate this moving forward is continuing to **phase out big cell towers and instead build lots of smaller towers**. • There would be benefit in **more focus on reopening waterways**, in addition to highways and local roads; they are a good alternative route/delivery system.

APPENDIX A

	- **Public-private partnerships are essential**, as resilience-building efforts (e.g., real estate developments, the construction of 5G cell towers, data sharing) more often than not fall outside the government's regulatory purview. **Incentivization of private efforts may have unintended consequences**, however, and can lead to government provision to private partners with little payoff. To avoid this pitfall, it is important to continually assess projects' effectiveness and hold partners to account. Relatedly, community engagement including grassroots stakeholders can sometimes provide a natural counterbalance. - **Community engagement and buy-in is important for projects to be successful**, but it also requires incentives, especially for community members with limited time and resources. This includes funding projects that help communities in blue skies environment too, not only during a disaster, so communities see multiple benefits. - **Pre-event prioritization of projects is helpful** to improve the allocation of limited resources available for investment after an emergency event.
Project Prioritization (Prioritization Exercise)	Participants commented: - **Survivability is still a priority**—people need their basic needs fulfilled, such as ensuring access to water. This also includes the ability to communicate with close connections, meaning friends and family. It would be useful if projects then moved quickly from restoring stability to improving the current conditions. - Resilience is a mindset and methodology. **Integrating resilience criteria into all aspects of day-to-day operations** is an effective option to achieve resilience in the long term. Currently, it is relatively easy to get funds for rebuilding efforts, but harder to get funds for preparedness, general upkeep, or any of the "not exciting" routine needs that often get overlooked but are required for achieving long-term resilience; these items must be funded. - **Resilience also includes focusing on the function of a system**, and investments should support the system approach more. - **Ensuring equity is a matter of both process and outcomes**, two prongs that are self-reinforcing. Community engagement is essential but insufficient to ensure the ultimate outcome is equitable. Equity must be deliberately considered throughout project design and implementation. **Relatedly, community engagement needs to integrate public education campaigns**, extending beyond formal, easy-to-reach organizations (churches, schools, etc.) to include harder-to-reach individuals. - **Reliance on private wells**, especially in rural areas, **would benefit from more attention**—monitoring, testing, repairing. Even though this type of project did not ultimately fall into the highest priority bin of projects during this workshop, there was significant discussion about this topic, as it impacted key priorities such as the community, vulnerable populations, and the environment.
	Protracted Oil Spill Scenario
Project Brainstorming (P&P Exercise)	Participants commented: - **Make the Oil Spill Liability Trust Fund more effective and impactful or consider a new and updated version of it**. It would be beneficial to revisit and update the conditions for drawing upon the fund, in advance of an emergency. It would also be beneficial to consider realistic triggers to access funds and to consider administration by a third party, including a committee, nonprofit, or

	academic institute, through a trust. It is also important to remember that a single fund can run out during a catastrophic spill, so a comprehensive program with alternative funding sources too—grants, revolving funds, low-interest loans—would be useful. • More planning and coordination is needed to **set up a dedicated Emergency Operations Center (EOC) or incident command and staging areas for a major spill**. This requires large, dedicated land (e.g., 500 acres) in an area that allows equipment to be protected from weather and that has multiple access points. • There was attention to but also disagreement about **burying flow lines, pipelines, and fiberoptic cables**—and rock dumping on flow lines. Some participants were very much in favor of these actions, and others believe it is just not realistic because currents in the Gulf expose buried pipelines eventually and scatter rocks placed on top, which then have to be picked up to comply with regulations stating zero equipment can be left on the ocean floor. Also, it is very difficult to inspect a buried pipeline. All agreed more research and R&D would be useful. • **Prioritizing "do no harm" when projects are planned** is more effective than repairing damage after the fact. • Much emergency response planning is for severe events, but **most will be low to medium in scale**, so some projects should focus there as well.
Project Prioritization (Prioritization Exercise)	Participants commented: • **It would be beneficial to place focus on prevention of the spill in the first place.** • **Resilience includes hardening infrastructure, mitigating impacts, and bouncing back from impacts**, but it also includes maintaining utilities—power and water—which are essential to maintaining resilience. • **After we ensure our survival, we should focus on an increased quality of life.** Although a sense of normalcy brings calm, disasters provide an opportunity to improve on the normal, so there is a need to return to stability followed by improvement. A response can promote equity and differentiate between a return to normalcy and effective recovery. • **Projects must be led locally whenever possible, to be effective.** All stakeholders should engage the whole community in efforts, provide opportunities for engagement, and not speak for anyone. They should support a public understanding of project implementation—explain what is happening and how money is spent. • **Quantitative measures are good, but it is also important to be wary of what we are counting** and make sure those things matter. • **Information sharing is important**, especially as grant-funded projects and environmental projects tend to get siloed and not understood as part of a bigger system or dataset.

APPENDIX A

TABLE A-4 Key Takeaways by Domain, by Scenario, and by Exercise: Society's Needs

	Society's Needs
\multicolumn{2}{c}{**Hurricane Scenario**}	
Project Brainstorming (P&P Exercise)	Participants commented: • **It is important to address problems that are well known and long-standing** (e.g., residential power loss, damage from storm surge, long-term economic hardship) but that have still not been accomplished. The reasons that these problems were never fully addressed have included lack of funds, awareness, attention, vision, knowledge, or staffing. This could make a new influx of federal support vital to new projects' success. • **A focus on housing, and returning to pre-hurricane habitability, is important.** Housing is the basis for a functioning economy, environment, and society. Projects that either enable residents to return to normal or make their housing more resilient ahead of a storm, through resilient building codes or stronger building materials, are beneficial to many aspects of society. Housing projects should emphasize incentives for designing, testing, marketing, and building new construction (and making repairs) with sturdier, more durable, more environmentally friendly materials. • **Investing in infrastructure with resilience and inclusiveness in mind necessitates understanding who the population is and what its needs are.** Health and housing projects are particularly sensitive to the needs of traditionally underserved, high-vulnerability populations. It would be helpful if projects increased this understanding. • **Enhance individual, small business, and local community resilience**, reducing the reliance on state and federal resources to respond and recover to an event.
Project Prioritization (Prioritization Exercise)	Participants commented: • **Impacting multiple prioritization criteria** in the same project is important. The value of a project is how holistically it will improve conditions (e.g., better or updated flood-mapping analysis). • **Community involvement** in a project is a priority. This includes a need to garner buy-in from the local communities who hear of grand projects that never come, or only understand the negative aspects of projects that have been implemented around them. • Prioritization criteria related to **project governance** were very clear-cut. It usually came down to elements that support a **coordinated vision** for infrastructure improvement and bettering peoples' lives, through **strong, effective leadership**. Participants noted that there is a lot of uncoordinated effort going on now that is not as helpful as it could be if it were better coordinated.
\multicolumn{2}{c}{**Protracted Oil Spill Scenario**}	
Project Brainstorming (P&P Exercise)	Participants commented: • There is concern with understanding **baseline data** in the Gulf region and how protracted oil spills might or might not affect the baseline. This includes data on direct impacts from protracted oil spills to beaches and aquatic environments near

	shore, marine ecosystem impacts of slow-release oil spills in an area that already has natural oil seepage, and health and human safety from cleanup operations. • There is a **lack of current baseline health and economic data, particularly among socially vulnerable populations**. Projects should collect and establish data against which data from a spill event can be compared to inform response and recovery decision-making. • **Employ effective public information campaigns** to combat mis- and disinformation, build public trust and engagement, and encourage revenue-generating activities (e.g., tourism, seafood consumption), all of which contribute to community and business resilience in the wake of the spill. Improve communications and messaging to the local population as well as tourists and visitors. An evidence-based communications and messaging operation run by trusted actors is critical to managing oil spill fallout. • Promote **opportunities for workers to be retrained** in other industries or provide immediate funding to affected workers and small businesses. A variety of funding mechanisms can be employed to put these systems in place before another oil spill occurs.
Project Prioritization (Prioritization Exercise)	Participants commented: • **Clear and coordinated communication including trusted members of society** is important. Informing the community and stakeholders of planning decisions and eliciting their feedback is important for projects to be successful. • **Increasing understanding about protracted oil spills**, particularly how oil spills impact the local population, is needed. Gaining a better understanding of actual impacts will give local authorities more information about how to mitigate those impacts. • **Supporting business continuity** and programs that will help keep businesses (and employment, tax base, tourism) afloat until recovery from the oil spill (or other impact) is complete, should be a focus area. • **Supporting worker economic mobility** by improving training and education programs that will allow workers to learn skills in different industries would be useful. Advocating for public-private partnerships in retraining before an oil spill happens so that when it does, the system is already set up to take in affected workers, would also be helpful.

TABLE A-5 Key Takeaways by Domain, by Scenario, and by Exercise: Environmental Protection

Environmental Protection	
Hurricane Scenario	
Project Brainstorming (P&P Exercise)	Participants commented: • Focus on projects that **protect, preserve, and restore the environment** from natural disasters as well as account for the impacts of climate change. • Robustness is multifaceted and can be achieved in different ways, ranging from **building coastal barriers** to **changing zoning laws for petrochemical plants** to address different aspects of infrastructure resilience. They are all important.

APPENDIX A

	• **Data and information sharing should be encouraged and improved** among organizations and with the community. Currently, many datasets that can benefit preparedness, mitigation, and/or response exist in silos, are poorly socialized, or simply are not leveraged in decision-making or in communications with the public. This problem is further exacerbated because the coast is under many agencies' jurisdiction; for example, NOAA (National Oceanic and Atmospheric Administration) conducts sea-floor mapping, Port Authorities monitor the ports, and the USACE (U.S. Army Corps of Engineers) and U.S. Coast Guard monitor other sections of the Gulf. • **Increased baseline data and monitoring would be helpful for detecting changes and impacts** caused by hurricanes, sea level rise, and other hazards. Much of this data must be developed. • While long-term restoration projects are central to environmental protection, **meeting the community's basic needs such as clean water in the aftermath of a natural disaster is very important**. Safeguarding drinking water and wastewater infrastructure and carrying out decontamination projects contributes to short-term community resilience.
Project Prioritization (Prioritization Exercise)	Participants commented: • Projects that would have **positive long-term impacts on the well-being of communities and ecosystems** would be useful. Projects that focus on prevention or avoidance of downstream harm and reduce the need for additional, response-type projects should be prioritized. • The "most-bang-for-your-buck" thinking would be beneficial, highlighting **projects that address multiple stressors**. Priority should be given to projects that would prevent more than one chain of adverse events from happening in the first place. • There are additional specific characteristics that are beneficial for resilience projects, including **addressing climate change, encouraging collaboration, leveraging existing data, maximizing positive impact on society, and promoting infrastructure systems/ecosystems to not just recover but also "spring forward."** • While doing no harm or reducing harm to the environment is important, it is also important to **think bigger and work to improve the environment**. It is also key to consider that there might be projects that could do some harm to the environment in the short term but have a greater benefit to the environment in the long term. • It would be useful to put more emphasis on **research into river and coastal erosion**.
	Protracted Oil Spill Scenario
Project Brainstorming (P&P Exercise)	Participants commented: • **It would be useful if corporations were held responsible** for environmental degradation and societal harm resulting from accidents of all scales. Consider issuing larger fines, requiring environmental offset projects, and requiring the establishment of funds for fisheries.

	• With the bathymetric data now available, **areas vulnerable to submarine landslides or other accidents are known. They should not be auctioned off for offshore drilling.** It would be helpful if the BOEM (Bureau of Ocean Energy Management) considered leasing standards and requirements that included provisions that would ban the abandonment of offshore infrastructure and make the organization responsible for their assets' maintenance from cradle to grave. • **Legal changes** would help increase enforcement and accountability. • It would be useful to support prevention and early detection of hazards through **enhancing and strengthening existing infrastructure (e.g., add sensors to pipelines to detect microfractures)** and **leveraging data from agencies like NOAA and NASA to inform decision-making**. • Investments in more **R&D into environmentally friendly response methods and tools** (e.g., biodegradable surfactants and dissolvents) could help minimize the negative impacts of oil spills on the environment. • Investments in **baseline studies on current or natural environmental conditions** are lacking and prevent comprehensive assessment of an oil spill's impact. Long-term epidemiology studies to understand impacts on human health are also lacking.
Project Prioritization (Prioritization Exercise)	Participants commented: • It would be useful to emphasize pragmatic and mitigation-focused projects. Given the scale of both active and abandoned offshore petroleum in the Gulf, it might be impossible to prevent oil spills from ever happening. Therefore, **focus on making ecosystems and restoration projects resilient to oil spills** when and if they occur. • Given the multiple stakeholders involved in oil spill detection and response, projects that **improve coordination, collaboration, and information sharing** among the relevant parties and with the public would be useful. • Projects that **address multiple stressors, reduce downstream costs, and contribute to long-term well-being of ecosystems and communities** would be useful. • "Hardening of Infrastructure" should include not just gray infrastructure but also **green infrastructure**, and it is useful to think of them together as a whole system. • Community members and local organizations can use techniques such as **crowdsourcing and geotagging** to conduct data collection on areas and species impacted by oil spills or tar balls.

B

Complete List of Project Ideas Identified

Tables B-1 through B-4 include the comprehensive list of project ideas that were brainstormed during Days 1 and 2 of the workshop. Projects in boldface are those that appeared in the high-priority bin listed in Part 1 of this proceedings. These are lists of project ideas that would contribute to increased infrastructure resilience in the Gulf region, to drive further thought and attention. This should not be interpreted as a definitive or authoritative list of projects.

TABLE B-1 Comprehensive List of Project Ideas Brainstormed during Days 1 and 2: Petrochemical Industry Functions

Petrochemical Industry Functions
Hurricane Scenario
• **Advance coordination of national power restoration priorities**
• **Bury power lines**
• Conduct remote inspections (drone, satellite)
• Create and streamline drone inspection protocols
• Develop more Commercial Driver's License (CDL) education programs to increase the number of truck drivers and lessen supply chain disruptions
• Develop short-term waiver programs for CDL hours
• Elevate backup power
• Elevate or protect chemical processing units required for refining
• Identify ways to protect tankers not able to evacuate ship channels
• Import electric power
• Increase flood mitigation at pump stations
• Increase industry-to-government coordination to enhance response
• Increase intragovernmental coordination to enhance response (federal-federal, federal-state)
• Increase resiliency requirements for storage tanks
• Increase secondary containment requirement for fuel tanks
• Install permanent and temporary flood protection
• Invest in smart tanks

- **Pass regulations establishing environmental, social, and governance (ESG) mandate for Corporations**
- Preauthorize an authority to clear waterways and open ports
- Preauthorize release from the Strategic Petroleum Reserve
- Preauthorize USACE (U.S. Army Corps of Engineers) and USCG (U.S. Coast Guard) authorities to reopen ports
- Preplan how to access gasoline sitting at refineries for local use
- Prestage generators and resources
- Prioritize fuel for first responders
- Provide access to nitrogen when shutting down refineries
- **Require and incentivize backup power for high-priority fuel stations**
- **Treat fuel stations as anchor institutions, prioritizing them for power restoration**
- Update designs for EV (electric vehicle) charging stations to protect against saltwater intrusion
- **Use flood sensor networks to map flooded areas**
- Use solar power at gas stations

Protracted Oil Spill Scenario

- Augment BSEE (Bureau of Safety and Environmental Enforcement) to play a bigger role in enforcement of regulations
- Build a new skimming test facility
- Clearly define and document roles, responsibilities, and authorities of all parties
- Conduct a study on oil spill response best practices in other countries
- Conduct GIS (geographic information system) mapping of pipeline systems
- Conduct R&D into and testing of containment domes and systems
- Conduct R&D into assessment of risk in disturbing sediments

Petrochemical Industry Functions

- Conduct R&D into improved ability to assess volume and impact of spill
- Conduct R&D into skimmers in deep water and heavy seas
- Conduct R&D into submarine landslides risk
- Conduct R&D into subsurface safety valves
- Conduct R&D into the microbe breakdown of oil
- Conduct R&D into water column absorbent materials
- Conduct tabletop exercises
- **Develop better, faster public information-sharing campaigns to improve public trust**

APPENDIX B

• Develop sensor technology for oil spill detection and conduct related data analysis
• Enhance how the Ports work with industry during recovery
• Establish multi-company operations technology review boards
• Examine the techniques used for containment in other spills
• **Expand and/or refine the mandated cleanup fund paid for by the petroleum industry**
• Harden blowout preventers
• **Implement stricter industry owner permitting requirements**
• Improve lessons sharing among oil industry companies
• Improve oversight of aging and abandoned oil infrastructure
• **Increase local leadership in response**
• Invest in advanced technology for containment, plugging
• Invest in better debris removal systems subsea
• Invest in better platform and anchor design (increase security, redundancy)
• Invest in Joint Industry Party (JIP) in oil infrastructure integrity technologies
• Invest in new pipeline construction technology to build resilience
• Invest in predictive modeling and forecast movement of oil and dispersants
• Invest in relief well technology
• Model seismic activity to predict where slides might occur; coordinate with the mining industry as appropriate
• Put requirements for containment plans and response in place before permitting
• Require the responsible party to disclose operational data for 6 months prior to spill
Petrochemical Industry Functions
• Review all related authorities and change laws as needed
• Review legacy permits and require containment plans where none are in place

TABLE B-2 Comprehensive List of Project Ideas Brainstormed during Days 1 and 2: Other Infrastructure Functions

Other Infrastructure Functions
Hurricane Scenario
• Assess and improve prestaged response assets
• Assess building codes in light of climate change
• Assess compound and cascading impacts of hurricanes on infrastructure

- Build in adaptability during infrastructure repairs
- Conduct remote damage assessment, for example, via drones
- Conduct satellite sweeps before and after storms to support damage assessment
- Create and enhance programs to monitor and test private wells for contamination
- Create economic incentives for cell providers to harden and install generators
- Deploy and test backup generators to support the water system
- Develop a new hoteling system for responders
- Develop and compile nationwide infrastructure best practices and make them available
- Develop and improve systems for soil water/moisture monitoring
- Develop better and more flood monitoring of low-lying roads and railways
- Develop designs for electricity infrastructure to withstand high winds pre-disaster, so designs are ready when the time comes to replace or build them
- Develop forecasts for impacts and damage
- Develop innovative ways to design transportation systems to connect people to societal needs
- Develop more effective and efficient ways to remove waterway obstructions from ports
- Develop new, more, and better flood and flash flood warning systems
- Develop or support programs for rapid deployment of storm surge detectors, pre-permit them and preplan the logistics
- Ensure cell tower design can withstand greater than 150-mph winds
- Ensure that fiber optic cables run to critical community-based facilities
- Evaluate if and when satellite communications are appropriate alternatives
- Further develop an alternate renewable power source, for example, solar
- **Harden, bury, or elevate electrical infrastructure (including distribution facilities, transmission structures, substations, and account for saltwater intrusion when burying them**
- **Identify and map infrastructure systems, and make this information available in a GIS (geographic information system) tool**
- Identify priority sites for communications restoration, just like jurisdictions do for power restoration
- Improve barge fleeting areas
- Improve warning systems to critical facilities
- Increase education and socialization on the coordination of response authorities
- Invest in more and new technology against saltwater intrusion into the water system
- Promote follow-up and socialization of the National Disaster Recovery Framework (NDRF)

APPENDIX B

- Promote keeping landlines for backup communications, and modernize them
- Promote wireless sensors and staff gauges to measure storm surge
- **Protect, harden, and/or elevate water infrastructure**
- Reevaluate and better communicate evacuation plans with clear destinations of where people should go
- Separate the internet from its current power source
- Set aside fuel for first responders and hospitals
- Study impacts of sea level rise (SLR) and climate change on the water system
- Support interoperability between wireless providers
- Support lowland land grading, especially in vulnerable communities
- **Transition to 5G with hardened towers and battery backups**
- Trim trees around electrical wires and water pipes
- Use more sensors and other monitoring systems for detecting chemicals in water

Protracted Oil Spill Scenario

- **Conduct a community needs assessment before cutting services in the wake of declining budgets**
- Conduct a ports and waterway safety assessment, including appropriate ship rerouting
- Conduct R&D into better and more reliable communications and navigation systems
- Co-share investments including Operations and Maintenance (O&M) to build up port infrastructure
- Develop a better system to evaluate claims for supporting funds
- Develop a bigger and better oil company–funded relief fund with a third-party administrator
- Develop a call center for offshore information (similar to 311)
- **Develop a data center for citizen and community data**
- Develop a new Emergency Operations Center (EOC) and Logistical Staging Area (LSA) with a predesignated, protected, and accessible location
- **Develop a Strategic Water Reserve**
- Develop alternative power sources to natural gas
- Develop a public education/emergency preparedness campaign, including likely impacts to infrastructure and what actions individual should take, when, and how
- Develop or enhance federal assistance programs for local government for infrastructure support in the wake of an oil disaster
- Develop renewable power sources for marginalized groups
- Expand air traffic control offshore in the Gulf region

• **Expand broadband access, starting with coastal areas**
• Expand the oil company–funded trust fund into a diverse portfolio of support
• Fund or partner to develop programs that provide for in-person/hard-copy assistance to residents for program registration, in addition to online registration
• Improve the oil company–funded trust fund with better triggers and management
• Incorporate long-term spills into emergency plans, especially regional transportation and economic development plans
• Increase capacity to scale up communications system support when impacted
• Increase environmental intelligence, particularly more ocean observation to support prediction of landslides and oil spills
• Increase transparency and information sharing with the media in advance of oil spills
• **Invest in R&D for oil spill barrier construction**
• Protect fiber and critical flow lines—bury them or use enhanced rock-dumping methods
• Research and plan better ways to back up communications infrastructure and scale up that response
• Streamline existing grant programs that support residents
• **Study impacts of climate change on infrastructure**

TABLE B-3 Comprehensive List of Project Ideas Brainstormed during Days 1 and 2: Society's Needs

Society's Needs
Hurricane Scenarios
• Account for the needs of nomadic populations
• Better socialize waterway closures
• Build capacity pre-disaster for business continuity planning, especially for small businesses
• Check shelter HVAC (heating, ventilation, and air conditioning) systems
• Create incentives for moving out of high-risk areas, including expanding related mitigation grants
• Create partnerships between government, community organizations, and industry to develop multipurpose affordable and safe temporary housing solutions
• Develop a digital home-to-community network
• Develop additional types of "remote" learning options
• Develop alternative designs and technologies for anticipating and addressing a foot of water in buildings
• Develop better plans to transition to remote learning

APPENDIX B

- Develop disaster education for high schools
- **Develop incentives for green technology and disaster-resistant building materials**
- Develop more enforceable liabilities and penalties
- Develop programs to prevent small businesses from closing
- Enhance distribution of supplies and potable water distribution
- Enhance erosion monitoring
- Enhance flood and other network sensors to support safe transportation
- Enhance hospital evacuation plans, including the transfer of patients prior to storm arrival
- Enhance levee motion sensors
- Enhance nursing home evacuation plans
- Enhance STEM (science, technology, engineering, and mathematics) programs
- Evaluate and enforce mold codes
- Facilitate the completion of applications for assistance by community members
- **Fund and overhaul Federal Emergency Management Agency (FEMA) flood maps for long-term planning in a changing environment**
- Identify and monitor vulnerable sewage lines to prioritize maintenance and repair
- Improve coordination through the National Business Emergency Operations Center
- Improve flood-mapping technology
- Improve neighborhood clinic operability during and after events
- Improve retention system for flood waters
- **Incorporate community engagement into planning for communities and schools**
- Monitor roads and supply lines around schools
- Phase out grandfathered houses in flood zones
- Plan for blood, critical medical consumables, and storage requirements
- Plan for potable water for dialysis centers
- Plan for redundant power, for example, generator maintenance, for community support sites
- Preidentify fleet asset management locations
- Preidentify logistics hubs on high ground
- Prepare mobile health-care clinics with locations identified for both fair skies and disasters
- Procure fuel trucks to support moving people
- **Rebuild and repair transportation infrastructure anticipating future floods**
- Redefine "critical infrastructure" to include schools

- **Reevaluate and streamline FEMA processes in general, for example, housing, flood insurance, incentives**
- **Regrade roads with more adaptive materials, for example, porous materials, living streets**
- Reform National Flood Insurance Program and require everyone to purchase insurance
- Review the safety of shelter locations
- Safeguard supplies of diapers and formula
- Study and improve private-sector supply chains
- Support cost-share management for grants
- Utilize green technology to support stormwater management

Protracted Oil Spill Scenario

- Assess the impacts of landslides
- **Conduct baseline studies for community health and needs, especially in underserved communities**
- Conduct exercises with an oil spill scenario with health-care facilities and the community to
- support training to manage expectations
- Conduct long-term studies on the impacts of oil spills, including comparisons of protracted oil spills versus those similar to Deep Water Horizon
- Conduct research into the impact of oil spills on the local population
- Conduct research into the impact of oil spills on the shore
- Conduct R&D into containment technology
- **Conduct research on alternative oil response technologies to prevent and/or mitigate adverse impacts from spills**
- Conduct R&D on the impacts of spill size on the environment
- Conduct research on fatigue and the safety of different shift lengths
- **Create and expand programs to support local industries after disasters**
- Create baseline studies of how historical events impacted the economy
- Decommission risky wells
- Develop a better system to enable teachers to have competitive pay and education and economic mobility
- Develop a mechanism for systematic research funding
- **Develop and implement training and education programs to improve economic mobility**
- Develop fishermen fleets as assets for ecosystem monitoring
- **Develop plans for the communication of science, including messaging to and awareness for the public**

APPENDIX B

- Develop programs that support estuary and wetland preservation and construction
- Develop programs to support beach communities intimately connected to fishing
- Develop public health monitoring programs ready for immediate deployment at oil spill time
- Develop trusted information sources with the community prior to disasters
- Ensure availability of safety equipment
- Establish co-ops to pool and negotiate for fuel and supplies
- Evaluate the Ocean Energy Safety Institute 2.0 program and its benefits
- Fund existing and new ecosystem monitoring and detection projects
- Have the CISA (Cybersecurity and Infrastructure Security Agency) monitor underwater pipelines
- Identify the needs of traditionally underserved communities
- Improve cleanup coordination, make roles and responsibilities clear
- Improve response techniques to reduce fish habitat damage
- Increase executive liability for spills
- Increase situational awareness for pipelines
- Increase transparency of communications
- Increase visibility (via websites and media campaigns) on cleanup and restoration activities
- Maintain active, continual communication programs with the community
- Make mental health services standard for our health systems
- Provide Wi-Fi and communication hubs for the community
- Regulate shift lengths
- Research and review the lawsuit injunction process for new projects, develop new standards
- Research compound and cascading impacts of oil spills
- Sheen boats in the Houston ship channel
- Strengthen the public information system and campaigns
- Study the risks of protracted spills versus large and fast spills
- Supply battery chargers for personal calls
- Sustain funding for sea grants
- Use sensors for rapid assessment of oil chemicals

TABLE B-4 Comprehensive List of Project Ideas Brainstormed during Days 1 and 2: Environmental Protection

Environmental Protection
Hurricane Scenarios
• Assess small water management system vulnerabilities
• Assess saltwater impact on concrete and steel
• Assess, monitor, and report long- and short-term impacts
• **Build and protect natural protection such as coastal systems, reefs, wetlands, and sea grasses**
• Build barriers around chemical facilities
• **Change construction standards beyond the 100-year flood**
• **Conduct rigorous epidemiological studies**
• Create a sensor network to monitor pollutants
• Create or enhance debris management plans
• Create or enhance flood protections around superfund sites
• **Design and implement restoration projects for wetlands and disturbed habitats**
• Develop a risk matrix, including trigger points for requesting and receiving waivers
• Develop alternative first-flood building designs for existing buildings, such as critical infrastructure and private homes
• Develop and enhance early warning systems for industry
• Develop and enhance saltwater barriers in coastal areas to combat flooding
• Develop mechanisms to monitor grids autonomously
• Ensure a consistent supply of chlorine
• Ensure availability and readiness of sewage overflow post-flooding testing units
• Improve access to geospatial data, including the creation of a shared GIS (geographical information system) platform and better coordination of maps with the EPA (Environmental Protection Agency), FEMA (Federal Emergency Management Agency), and other agencies
• Improve damage assessments by using LIDAR flyovers and 3-D cameras to estimate debris
• **Improve geospatial data use and sharing for mitigation and early response**
• Improve geospatial data, including coastal risk inundation maps, flash flood monitoring, digital elevations, and commercial data (high resolution)
• Incentivize industry to prioritize ESG (environmental, social, and governance) framework
• Increase USCG (U.S. Coast Guard) inspection of regulated facilities
• Invest in infrastructure that supports post-disaster debris management

- Invest in natural (green) infrastructure solutions near shore
- Invest in secondary containment requirements to capture chemicals
- Maximize beneficial use of dredge material
- Develop protection systems for wastewater plants against flooding
- Require continuous monitoring and reporting of pollutants
- Research better mechanisms for transport of water during flood events
- Research point source and non-point source pollutants and their intersections with vulnerable populations
- Revisit zoning for chemical plants
- Store potable water close to populations
- Study long-term impacts on living organisms

Protracted Oil Spill Scenario

- Add greenhouse gas (GHG) emissions penalties to EPA (Environmental Protection Agency) spill regulations and permitting
- Advance and facilitate the inclusion of citizen science and observations
- Build oil spill resiliency into restoration projects
- **Change BOEM (Bureau of Ocean Energy Management) leasing to include cradle-to-grave infrastructure and account for equity**
- Collect more and better bathymetric information
- **Conduct epidemiological health studies to assess chronic impacts of pollutants on coastal communities**
- Conduct R&D on dispersants early use and review the current authorization for use
- Conduct R&D on stronger tar ball intervention, detection, and notification, and cleanup Technology
- Conduct risk assessments for platform risk and mitigation; use them to inform foundation and anchor design criteria
- Conduct surveys of wildlife habitats
- Conduct underwater pipeline monitoring for leaks and microfractures
- Create numerical, model-based forecasts for landslides
- Develop alternative fishing opportunities, for example, oyster aquaculture
- Develop and enhance automatic identification system (AIS) and predictive routing for vessel tracking and movement
- Develop and enhance methods and tools to measure hydrodynamics and spill rates
- Develop measures to track recovery
- Develop mechanics for oceanic continuous monitoring, including oil monitoring

• **Develop projects to improve the health of ecosystems to make more resilient coasts and barriers (e.g., land conservation, storm water management, EPA Best Management Practices)**
• Evaluate the scope of the mudslide threat
• Expand use of NOAA (National Oceanic and Atmospheric Administration) fish management processes
• Identify and monitor natural seeps
• Identify endangered species and update regulatory requirements
• Improve water column oil absorbents
• Improve subsea well caps
• Increase applications of remote sensing to monitor the scale of impacts on the air, land, and Water
• Increase collaboration among research agencies and add a budget line for this
• **Increase community resilience by providing mental and physical health support services**
• Invest in R&D product development of environmentally friendly biodegradable dissolvents and surfactants
• Leverage fish hatcheries to reintroduce healthy populations
• Locate an offshore command post to manage information and communications during a spill
• Put emission capture technology in place
• Reassess the fisheries management process to allow sufficient time for recovery
• Require the private sector to conduct offset projects
• Research the long-term fate and transport of dissolvents and surfactants
• Study climate-driven migration of habitats and conditions; account for this during restoration
• Understand and identify ways to mitigate impacts on tribes and Indigenous populations reliant on coastal flora and fauna for cultural practices
• Use existing vessels mounts for monitoring and to collect baseline data

C

Prioritization Framework: Research and Rationale

Eleanore Douglas, Ph.D.

This Appendix provides an overview of the research and rationale that drives the development of a prioritization framework for infrastructure resilience projects in the Gulf of Mexico region, advanced through this workshop. It starts with the literature review overview. Then it introduces three basic approaches to framing and orienting a framework for this purpose of project or investment prioritization, three options for supporting project prioritization, the six-step process proposed for this framework, and a closer look at Step 3 of the six-step process. Step 3 was the focus of the workshop because it was best positioned to offer the most value to other decision-making agencies, as this is an immediate and concrete challenge that many agencies are likely to face in the short term. The appendix concludes with potential next steps to continue building out this prioritization framework.

First, workshop designers surveyed general approaches to, and mechanisms that support, making investment decisions specifically in the public policy space. This literature review included previous National Academies of Sciences, Engineering, and Medicine studies and a number of formalized frameworks for resilient infrastructure decision-making published by other institutions over the past decade (see Box C-1).

The literature review revealed three general approaches to framing and orienting the authority and legitimacy of such decision-making: (1) top-down, (2) bottom-up, and (3) technocratic. Top-down approaches tend to align the legitimacy of resource decisions with strategic vision and values, planning and operational guidance, or other national or executive-level policy. They analyze proposed projects or investments in light of their explicit alignment with and support for these guidance and policy documents. The bottom-up approach draws its authority and legitimacy from the populations affected by the resource decisions being made. As a result, this approach tends to focus on representative participation or sampling of the relevant populations, as well as the accurate elicitation and reflection of the populations' preferences to help shape decisions on investments. The technocratic approach tends to source the authority and legitimacy for its decision-making in explicit scientific, expert, and technocratic analysis and formalized understanding of the space in which investment decisions are being considered.

While there are currently government entities and other organizations that adhere to each of these approaches (e.g., Department of Defense as "top-down," local school-boards and city councils as "bottom-up," and the Environmental Protection Agency as "technocratic"), most public policy decision-making constructs tend to reflect a mixture of all three approaches. Nevertheless, decision makers must understand the different sources of authority and legitimacy underpinning their decisions, in order to convey their decisions effectively across a population, whose perceptions of authority and legitimacy may differ significantly.

> **BOX C-1**
> **Key Sources in Literature Review to Support Development of the Prioritization Framework**
>
> **National Academies Studies:**
>
> - Building and Measuring Community Resilience: Actions for Communities and the Gulf Research Program
> - Investing in Transportation Resilience: A Framework for Informed Choices
>
> **Other Institutional Frameworks:**
>
> - The City Resilience Framework by the Rockefeller Foundation and ARUP
> - Global Resilient Cities Network's Toolkit for a Resilient Recovery
> - Zurich Flood Resilience Alliance Framework
> - GIB Foundation's SuRe® Standard for Sustainable and Resilient Infrastructure
> - DNV GL's Systems and Urban Resilience Framework (SURF)
>
> **Other U.S. Government Frameworks and Tools:**
>
> - National Institute of Standards and Technology (NIST) Community Resilience Economic Decision Guide
> - Federal Emergency Management Agency (FEMA)
> - U.S. Army Corps of Engineers (USACE)
> - U.S. Department of Housing and Urban Development (HUD)
>
> **Private-Sector Environmental, Social, and Governance (ESG) Investment**
>
> - Adasania Social Justice Investment Criteria

Several mechanisms exist for framing and providing analytical support to public investment decision-making that blend the three approaches. Given the mission of the National Academies and the Gulf Research Program (GRP), and the focus of this effort being infrastructure resilience during disasters, three options for supporting project prioritization emerged as most viable—Scenario-based Systems Analysis Construct, Modular Table-Top Game for Local Engagement, and Risk-based Portfolio Management Approach. Portfolio management was best suited to the task because of its flexibility on specificity and ability to address trade-offs, including across sectors. The other two work best with a narrow focus and narrow criteria, or are difficult to scale up to a regional level. As a result, we decided to use the risk-based portfolio management approach, drawing on elements of the others, using scenarios and serious games in support of that approach.

RISK-BASED PORTFOLIO MANAGEMENT APPROACH

The main objective of a portfolio management approach is to make the best trade-off of risk against return across a number of different investments or assets within the context of a

APPENDIX C

particular goal or criteria. Portfolio management tools and methods are powerful due to their explicit engagement with both risk and uncertainty and their inherent flexibility with respect to almost every other aspect of the analytical context. Portfolio approaches are neutral with respect to the organizational context and the particular selection of governing criteria. This has made it a useful framework for supporting investment decision-making across a broad range of different organizations and criteria, from private-sector financial investment management to the National Institutes of Health priority setting and grant management to the Department of Homeland Security R&D/Science and Technology (S&T) portfolios. A weakness of portfolio management, particularly in the context of infrastructure investment decision-making, is the tendency of traditional portfolio analysis tools to treat each project or proposed investment as a fully independent item, balancing them collectively only across the whole group. It takes additional layers of analysis to acknowledge and/or track interdependencies between particular projects or to evaluate sub-clusters of projects within a larger portfolio. Still, the risk-based Portfolio Management Approach lends itself most readily to the objectives of this National Academies effort and therefore serves as the general driver for the process developed and implemented during the workshop.

ELICITING STAKEHOLDER INPUTS

Considering existing frameworks for portfolio management in a public policy and infrastructure context and then adapting them to support the prioritization of projects likely to increase infrastructure resilience in the Gulf region, workshop designers proposed a six-step process for a useful framework (see Figure C-1).

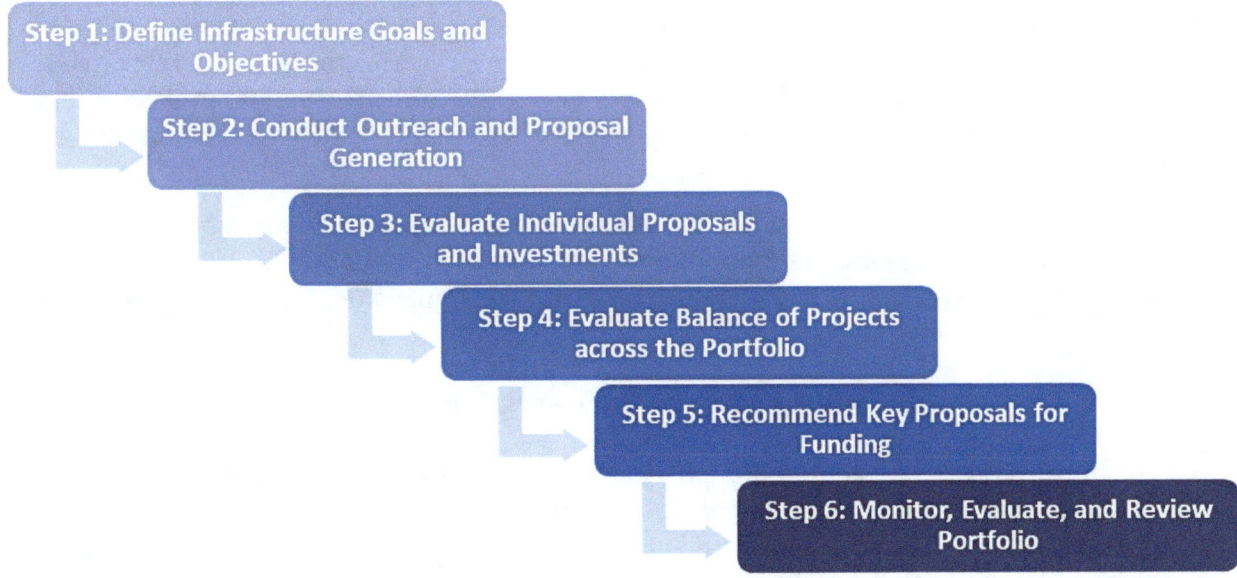

FIGURE C-1 Key steps to a framework that support prioritization of infrastructure resilience projects in the Gulf of Mexico region.

While the six steps provide a comprehensive approach to risk-based portfolio management, the National Academies determined that, to use the limited time of participants most effectively and develop actionable, practical outputs, this workshop would focus on refining, validating, and potentially ranking the criteria by which infrastructure project proposals can be effectively analyzed, or Step 3 of the process.

The literature review also revealed a number of key criteria useful in analyzing and evaluating the impacts of particular actions, projects, and investments with respect to resilience. Workshop designers leveraged the resilience and investment frameworks referenced above as well as current Federal Emergency Management Agency (FEMA), Environmental Protection Agency (EPA), and U.S. Army Corps of Engineers (USACE) grant evaluation guidelines, to capture a variety of potentially salient criteria. First, workshop designers grouped these criteria into five key macro-criteria categories: Environment, Economy, Society, Resilience, and Project Governance (see Figure C-2).

Environment	Economy	Society	Resilience	Project Governance

FIGURE C-2 Macro-criteria that support the prioritization framework.

The five macro-criteria served to help structure the literature- and participant-identified sub-criteria and prevent information overload during the workshop. Second, workshop designers reviewed the key criteria suggested by the literature and identified the subsets that would likely be relevant for this type of prioritization framework. Workshop facilitators reviewed this information to help drive discussion when ideas were not organically volunteered from workshop participants and for comparison purposes after initial brainstorming. The approximately six sub-criteria per macro-criterion that were common across all or most domains in the final set of master criteria developed during this workshop were also largely identified in this initial research. Several of the domain-specific sub-criteria suggested by participants were also identified in the initial research. While almost all of these sub-criteria emerged through small group discussion naturally and very limited leading on the part of the facilitators was necessary, there was still a great deal of consistency between sub-criteria suggested by participants and by the literature.

The National Academies and GRP knew that the workshop would not authoritatively validate a prioritized short list of sub-criteria; however, they did expect that useful and meaningful discussions would occur and some preliminary insights would be drawn, which was the case.

Building out the Framework Further

The macro-criteria and sub-criteria provide useful inputs into how Step 3 (and 4) of this portfolio management approach to a prioritization framework might be implemented. Step 3 focuses on evaluation of individual infrastructure project proposals and was the focus of this workshop. Step 4 focuses on evaluating infrastructure proposals as a group or a portfolio. Useful next steps apply to each of these steps.

Step 3

Lists of evaluation sub-criteria can be quite extensive. In fact, the set of criteria that was developed through this workshop includes between 12 and 14 sub-criteria for each macro-criterion. While it is important to be comprehensive in the consideration of each macro-category, for realistic implementation it is also vital to think about the two or three most indicative sub-criteria for each category. Further refinement is an appropriate next step.

As described in other sections of this proceedings, some participants felt that a couple of the criteria needed to permeate the entire decision-space and process more thoroughly than in just a single macro-criterion category. Examples include the macro-criterion of Resilience or the sub-criteria of Equity or Equity and Inclusion. Another next step would be to further explore if these criteria are useful at the project level, should be incorporated into portfolio-level considerations, or should be applied to both, and how.

Step 4

Another useful action to build out the framework further is to focus on the particular requirements of the portfolio evaluation step and explore what criteria are most important to balancing a portfolio of infrastructure investments, as opposed to making funding decisions at the individual project level. This will require a more explicit definition of infrastructure risk and what it might mean to buy down infrastructure risk. It would be useful for stakeholders to consider the most important sources of uncertainty in infrastructure investments. Important insights could be drawn from how existing federal and private-sector portfolio managers build and adapt their portfolios. Finally, the portfolio evaluation step is also the one at which systems-level analytical approaches might effectively be applied.

D

Event Agenda

Day 1: Hurricane Scenario
November 15, 2021, 9:30–4:30 EST
In-person at the Keck Center, Washington, DC

9:30-9:40	Plenary	Introductory Remarks • Lauren Alexander Augustine, National Academies of Sciences, Engineering, and Medicine • Micah Lowenthal, National Academies • David Daniel, University of Texas at Dallas (Emeritus) and GRP Division Committee
9:40–10:05	Plenary	Keynote • Jason Tama, National Security Council
10:05–10:30	Plenary	Keynote • Lt. General Thomas P. Bostick, Ret. US Army and Bostick Global Strategies
10:30–11:00	Plenary	Introductions and Roadmap for Day 1 • Erin Mohres, CAN
11:00–11:15		Break
11:15–12:40	Break-out Groups	Prevention & Preparedness (P&P) Exercise
12:40–1:40		Lunch
1:40–4:15	Break-out Groups	Prioritization Exercise
4:15–4:30	Plenary	Wrap up and Closing Remarks • Micah Lowenthal, National Academies

Day 2: Protracted Oil Spill Scenario
November 16, 2021, 9:30–4:00 EST
In-person at the Keck Center, Washington, DC

9:30–9:45	Plenary	Introductory Remarks • Lauren Alexander Augustine, National Academies • Micah Lowenthal, National Academies • David Daniel, University of Texas and GRP Division Committee
9:45–10:15	Plenary	Introductions and Roadmap for Day 2 • Erin Mohres, CAN
10:15–11:55	Break-out Groups	Prevention & Preparedness (P&P) Exercise • Rotate through two domains • 15-minute break • Rotate through two remaining domains
11:55–12:55		Lunch
12:55–3:30	Break-out Groups	• Prioritization Exercise • Rotate through two domains • 15-minute break • Rotate through two remaining domains
3:30–4:00	Plenary	Wrap up and Closing Remarks • Micah Lowenthal, National Academies

APPENDIX D 85

Day 3: Capstone
November 18, 2021, 9:30–4:30 EST
Virtual through assorted online tools

9:30–9:45	Plenary	Introductory Remarks • Lauren Alexander Augustine, National Academies • Micah Lowenthal, National Academies • David Daniel, University of Texas at Dallas (Emeritus) and GRP Division Committee
9:45–10:00	Plenary	Keynote • Marcia McNutt, President, National Academy of Sciences
10:00–10:30	Plenary	Introductions and Roadmap for Day 3 • Erin Mohres, Facilitator
10:30–11:00	Break-out Groups	Capstone Project Prioritization
11:00–12:30	Break-out Groups	Close Look: Prioritization Criteria Review • Review two categories • 15-minute break • Review three remaining categories
12:30–1:15		Lunch
1:15–2:15	Break-out Groups	Close Look: Prioritization Criteria Feasibility
2:15–3:00	Break-out Groups	Project Risk and Impact on Prioritization
3:00–3:15		Break
3:15–3:30	Plenary	Next Steps and Hotwash
3:30–4:00	Plenary	Wrap-up and Closing Remarks • Lauren Alexander Augustine, National Academies • Micah Lowenthal, National Academies • David Daniel, University of Texas at Dallas (Emeritus) and GRP Division Committee

E

Biographical Sketches of Speakers and Steering Committee

Keynote Speakers

Jason Tama, Director, Resilience and Response, National Security Council, The White House

Jason Tama is the deputy of the National Security Council's (NSC) Resilience and Response Directorate. In this role, Tama oversees national policy in a number of areas, including domestic preparedness and incident response, hazard mitigation and recovery, and critical infrastructure security and resilience. He is detailed to the NSC from the U.S. Coast Guard and most recently served as the captain of the Port of New York and New Jersey. Tama holds a bachelor of science in mechanical engineering from the United States Coast Guard Academy, a master of engineering from the University of California, Berkeley, and a master of business administration from the MIT Sloan School of Management. He is an MIT Sloan Fellow, Brookings Institution Federal Executive Fellow, Marshall Memorial Fellow, and White House Fellows National Finalist. While at Brookings, Tama completed extensive research on emerging national security threats, and his work appeared in multiple publications.

Lt. General Thomas P. Bostick, Retired, U.S. Army and Bostick Global Strategies

Lt. Gen. Bostick currently serves as the chairman of Bostick Global Strategies. Bostick recently served as chief operating officer and president of Intrexon Bioengineering (NASDAQ: XON). He was the 53rd chief of engineers and commanding general of the U.S. Army Corps of Engineers. Bostick helped lead the nation's response to Superstorm Sandy. He was the Army's director of personnel; he deployed with the 1st Cavalry Division during Operation Iraqi Freedom and later commanded the Gulf Region Division with responsibility for an $18 billion construction program. During 9/11, he was the senior watch officer in the Pentagon's National Military Command Center on the Joint Staff. He was an associate professor of mechanical engineering at West Point. As a White House Fellow, he was a special assistant to the secretary of veterans affairs. A member of the National Academy of Engineering, Bostick is a licensed professional engineer and a Forbes Contributor. He is on the boards of CSX (NASDAQ: CSX), Perma-Fix (NASDAQ: PESI), HireVue, American Corporate Partners (ACP), and Streamside Systems. Bostick is a graduate of the U.S. Military Academy, holds master of science degrees in civil and mechanical engineering from Stanford University, and a Ph.D. in systems engineering from George Washington University.

Marcia McNutt, President, National Academy of Sciences

Dr. McNutt is a geophysicist and the 22nd president of the National Academy of Sciences. From 2013 to 2016, she was editor-in-chief of Science journals. She was director of the U.S. Geological Survey from 2009 to 2013, during which time USGS responded to a number of major disasters, including the Deepwater Horizon oil spill. For her work to help contain that spill, McNutt was awarded the U.S. Coast Guard's Meritorious Service Medal. She is a fellow of the American Geophysical Union (AGU), Geological Society of America, American Association for the Advancement of Science, and the International Association of Geodesy. Her honors include membership in the American Philosophical Society and the American Academy of Arts and Sciences. In 1998, McNutt was awarded the AGU's Macelwane Medal for research accomplishments by a young scientist, and she received the Maurice Ewing Medal in 2007 for her contributions to deep-sea exploration. She received a B.A. in physics from Colorado College and her Ph.D. in earth sciences at the Scripps Institution of Oceanography. She holds honorary doctoral degrees from Colorado College, the University of Minnesota, Monmouth University, and the Colorado School of Mines.

Steering Committee Members

Dr. David Daniel, Chair, University of Texas at Dallas (Emeritus)

Dr. David Daniel is president emeritus of the University of Texas at Dallas. Previously, he was dean of engineering at the University of Illinois. Earlier, Daniel was L.B. Meaders Professor of Engineering at the University of Texas at Austin, where he taught for 15 years. Daniel has conducted research in the area of geoenvironmental engineering, including research on drilling fluids, containment and management of those fluids, and fluid pressure control in the subsurface. Daniel served as chair of the American Society of Civil Engineers' External Review Panel that evaluated the failure of the New Orleans levees caused by Hurricane Katrina. He also served as a member of the National Research Council's Nuclear and Radiation Studies Board, the Board on Energy and Environmental Systems, and the Geotechnical Board. Daniel received a Ph.D. in civil engineering from the University of Texas at Austin. He was elected to the National Academy of Engineering in 2000.

Prof. M. Granger Morgan, Carnegie Mellon University

Granger Morgan is the Hamerschlag University Professor of Engineering at Carnegie Mellon University. He holds appointments in three academic units: the Department of Engineering and Public Policy, the Department of Electrical and Computer Engineering, and the H. John Heinz III College. His research addresses problems in science, technology, and public policy with a particular focus on energy, environmental systems, climate change, and risk analysis. Much of his work has involved the development and demonstration of methods to characterize and treat uncertainty in quantitative policy analysis. At Carnegie Mellon he is co-director of the Center for Climate and Energy Decision Making and the Carnegie Mellon Electricity Industry Center. He has served on a number of advisory committees related to energy issues in the United States and Europe. At the National Academies, he was the National Academy of Sciences (NAS) co-chair of the Report Review Committee and has been involved in a variety of other NAS/National Research Council activities and studies. He is a member of the NAS, American Academy of Arts and Sciences, and a fellow of the Institute of Electrical and Electronics Engineers, the Society for Risk Analysis, and the American Association for the Advancement of Science. He holds a B.A. from Harvard College (1963) and a Ph.D. from the Department of Applied Physics and Information Sciences at the University of California, San Diego (1969).

APPENDIX E

Sara N. Ortwein, Exxon Mobile Corporation (Retired)

Sara Ortwein retired from ExxonMobil in March 2019. Prior to retiring, she was president of XTO Energy, a subsidiary of ExxonMobil, from November 2016 through February 2019 and was responsible for ExxonMobil's unconventional oil and gas business. Ortwein earned a bachelor of science degree in civil engineering at the University of Texas at Austin before joining Exxon Company, U.S.A., in 1980 as a drilling engineer. She held numerous technical and managerial assignments throughout her career. In 1997, she was named reservoir evaluation and planning manager for Exxon Ventures, CIS, focusing on new venture pursuit and capture in Russia, Azerbaijan, and Kazakhstan. In 2001, she became a corporate upstream advisor to senior management at ExxonMobil headquarters in Irving, Texas. Three years later, she was named production manager responsible for ExxonMobil-operated U.S. production operations. Ortwein was named vice president of engineering for ExxonMobil Development Company in 2006, where she was responsible for engineering design for major projects around the world. She served as president of ExxonMobil Upstream Research Company from September 2010 until November 2016, where she was responsible for research and technology development and application for ExxonMobil's Upstream business. In 2020, Ortwein was elected to the National Academy of Engineering. She currently serves on the Board of Directors of The Academy of Medicine, Engineering and Science of Texas. She is a member of the Society of Petroleum Engineers. She is past chair of the University of Texas Engineering Advisory Board and currently serves on the Executive Committee. In late 2020, she joined the Board of Directors of Sanara Medtech.

Lt. General Thomas P. Bostick (Retired, U.S. Army) is also a steering committee member.

His biographical sketch is included above under Keynote Speakers.

F

Event Participants

Juli Ansay, U.S. Army Corps of Engineers (USACE) Southwestern Division
William Ball, Sr., Southern Company Services (retired)
Phil Bedient, Rice University
Steward Behie, Texas A&M
Terry Boston, Terry Boston, LLC.
Brenda Breaux, New Orleans Redevelopment Authority
Jeff Byard, Team Rubicon
Jared Gartman, USACE Mississippi Valley Division
Angela Gladwell, Federal Emergency Management Agency
David Green, National Aeronautics and Space Administration
Phil Grossweiler, M&H Consulting
Liv Haselbach, Lamar University
David Haun, Haun Consulting, Inc.
Brian Hill, Maritime Administration
Maria Honeycutt, White House Office of Science and Technology Policy
Richard Hughes, Louisiana State University Petroleum Engineering
Veneeth Iyengar, Broadband Development and Connectivity, Louisiana State Government
Bruce Lambert, Maritime Administration
Kristin Ludwig, U.S. Geological Survey
Scott Lundgren, National Ocean Service, National Oceanic and Atmospheric Administration
John Lynk, Pipeline Research Council International, SolSpec
Brandi Martin, U.S. Department of Energy - CESER
Shubhra Misra, The Water Institute of the Gulf
Missy Partyka, Mississippi-Alabama Sea Grant
Jessica Rackley, National Governors Association
Danny Reible, Texas Tech University
Hanadi Rifai, University of Houston
Alan Robertson, Association of State Drinking Water Administrators
Daniel Schwanik, Maritime Administration
Buck Sutter, RESTORE Council
Kirsten Trego, U.S. Coast Guard
RADM James Watson (retired, USCG), American Bureau of Shipping
Charlie Williams, Center for Offshore Safety (retired)
Roy Wright, Insurance Institute for Business and Home Safety